T0135715

**BERGISCHE
UNIVERSITÄT
WUPPERTAL**

LOW-POWER OPTIMIZATION OF SELECTED FPGA BLOCKS

Dissertation

submitted at the deparment of
Electrical, Information and Media Engineering
University of Wuppertal

in fulfilment of the requirements for the degree of
doctor of engineering
(Dr.-Ing.)

Karol Niewiadomski, Dipl.-Ing.

2019

Referee: Prof. Dr.-Ing. Dietmar Tutsch
Co-Referee: Prof. Dr.-Ing. Carsten Gremzow (HTW Berlin)
Date of submission: January 22, 2019
Date of defense: May 10, 2019

Bibliografische Information der Deutschen Nationalbibliothek

Die Deutsche Nationalbibliothek verzeichnet diese Publikation in der
Deutschen Nationalbibliografie; detaillierte bibliografische Daten sind
im Internet über http://dnb.d-nb.de abrufbar.

ISBN 978-3-8325-4947-3

Logos Verlag Berlin GmbH
Comeniushof, Gubener Str. 47,
10243 Berlin
Tel.: +49 (0)30 42 85 10 90
Fax: +49 (0)30 42 85 10 92
INTERNET: https://www.logos-verlag.de

To my family.

For their support and never ending patience during countless hours.

Don't be encumbered by history. Go off and do something wonderful.

- Robert N. Noyce, Co-founder of Intel, the Mayor of Silicon Valley

Abstract

Reconfigurable logic can offer a wide range of flexibility for different applications. Since the introduction of the first FPGAs by a number of different companies, they were used as glue logic to connect different components of an entire system. In the following years, fast prototyping has led to new possibilities of firmware development running in parallel to IC design, and because of that, the development time has been reduced significantly and products can enter the market at an earlier point of time. However, these special ICs cannot compete with ASICs in mass production due to their high production costs which consequently limits their application to cost sensitive markets e.g. for being used in automotive applications. This has led various FPGA vendors to develop low-end FPGAs for countering the price advantage of ASICs and to raise the attractiveness of these chips in the automotive industry. Despite many advantages of reconfigurable logic, cost-optimized FPGAs lack of efficient power saving mechanisms which is a necessary feature in applications with limited energy resources.

The scope of this thesis is the research on the implementation of dedicated power saving logic on selected blocks within a FPGA. In order to do so, these blocks were first analyzed based on the choice of a commercial baseline architecture and re-engineered to significantly cut-down the average power consumption and related leakage currents. The affected blocks were identified as SRAM cells, D-FFs and I/O elements, which consume a comparably large area footprint of the FPGA fabric and therefore depict a good target for optimization and modification. Even though lowering the average power consumption was the primary goal of this thesis, further aspects of integrated circuits could not be totally neglected. These aspects are specific characteristics like SNM, WNM, maximum operating frequency, transistor count and high impedance ability and were also considered and evaluated against the baseline design as well as compared against selected solutions from related academic research. In addition to that, the newly developed D-FF's ability to counter security-related attacks is introduced in this thesis by referring to a special logic style. This special logic style hardens the D-FF against special attacks like DPA. All simulations performed for the purpose of this research are based on a $90nm$ process technology by TSMC as the baseline design is manufactured by using the same technology. The interaction of some of these cells is also investigated by integrating them to logic blocks of a higher hierarchical level. Almost all simulation results prove the paramount performance of the newly developed cells. These improvements come at the cost of a higher transistor count which subsequently leads to a higher area footprint after synthesis and therefore to higher costs. All applied measures used to extend battery lifetime demonstrate how to push low-power abilities in different designs to the limit without considering a penalty in costs. Thus, the selection of a special low-power FPGA for cost-sensitive vehicle applications must be carefully decided and clearly justified.

Contents

II

Chapter 1

Introduction

From the beginnings of history, mankind has been working towards the invention of methods and tools for overcoming daily challenges and improving living quality. One of those inventions of the last century was the creation of the junction transistor by William Shockley and further scientists, which was officially presented in 1951 [1]. Due to its tremendous impact on the development of electronic devices, this invention is highly regarded as the third industrial revolution. Notably, one of the most important benefits of this new technology was to have an alternative option to vacuum tubes, which were of high importance for various applications. However, vacuum tubes suffered from a number of drawbacks like size, power consumption and achievable lifetime. From that time on, an extraordinary development and progress in electronics and integrated circuits began. Whilst the dawn of discrete semiconductors like Light Emitting Diodes (LEDs), transistors and all imaginable deviations was certainly a remarkable step towards desirable characteristics in terms of sustainability, reliability and cost reduction, the upcoming of integrated circuits (ICs) was the real boost towards wide-spread availability of fast processing units for both professional usage and consumer market.

The first ICs that came out in the early 1960s, were used to implement simple functions like standard boolean operations in Dual-Inline-Package (DIP). It should be mentioned here, that the vast majority of these ICs were manufactured in silicon, even though it was not the best material choice in terms of performance. However, due to its availability and pricing, it has beaten any other technology currently available. The first implementations were done by using npn or pnp bipolar transistors, which are an inevitable device when it comes to the design of, e.g., amplifiers. Due to a high gain factor g_m of bipolar transistors, they are suitable to be used in operational amplifiers where high switching frequencies do not play an important role. However, it is possible to use bipolar transistors for implementing standard gates like AND, NAND, etc., but the circuit designers have to deal with some drawbacks. Due to a lack of alternatives, these gates were implemented in Transistor-Transistor-Logic (TTL) [106]. In many cases, and regardless whether npn or pnp transistors are used for implementation, these devices reach their limits in terms of maximum switching frequencies. Circuits implemented in TTL inherit the undesirable effect of higher power consumption based on their nature to work only when there is a constant current flow through the whole implementation. These restrictions led to the necessity of research for alternative designs, whose main characteristics would be an improved achievable operational frequency, better energy savings and a higher integration density. Research activities resulted in the invention of the Field Effect Transistor (FET) and its most commonly used derivation, the Metal Oxide Semiconductor Field Effect Transistor (MOSFET). This newly invented type outperformed its earlier predecessor in terms of switching speed and energy savings in a remarkable way. As counterpart to npn or pnp type biploar transistor, a major distinction was done by having n-channel (nMOS) and p-channel (pMOS) types, with different pros and cons. First circuits designs were engineered by using either nMOS or pMOS transistors by accepting the drawback of penalties in full swing voltage range at the output(s) of each gate.

Keeping in mind the above mentioned technological restrictions of the time, the most important logic style was introduced in the early 1970s: Complementary MOS logic (CMOS). Even today, many years after the dawn of the first CMOS designs, this logic style is still the backbone for the vast majority of ICs in commercial applications where special attention to some particular characteristics is not mandatory. It overcomes the disadvantages of TTL and combines beneficial abilities of nMOS and pMOS transistors. However, energy consumption did not play a significant role during research and development for a very long time. Other factors, e.g., integration density and speed were more important than power consumption, since mobile applications belonged to a negligible part of electronic devices. Increasing the yield in semiconductor manufacturing was (and still is) one of the most important holy grails of nearly all companies who were interested in decreasing their manufacturing costs for a higher revenue of each produced die. These proceedings in chip development and manufacturing have a significant influence on the daily life of everyone of us. It appears that microprocessors are involved in almost every aspect of life and are responsible for simplifying different things, which were cumbersome in the past [87]. The evolution of cars from simple road vehicles to rolling computers shows the impressive impact of progressing digitalization on applications, which were initially designed just for the sake of mobility. This original intention shifted over the years by a continuous improvement of safety towards autonomous cars without the necessity of meeting *driver-in-the-loop* requirements. Modern cars provide various Advanced Driver Assistance Systems (ADAS) to support drivers in daily driving situation, to increase safety and to provide a certain degree of luxury [89]. These features require, depending on the respective application, more or less computation power and sufficient energy resources. Here it comes to the point, where the rising number of Electronic Control Units (ECUs) challenges the limited battery capacities of road vehicles and therefore leads to necessary re-thinking steps of conceptual design. Since modern cars can contain more than 70 ECUs for the purpose of different supporting vehicle functions, the question for combining different functions on one ECU arises [89]. Adaptive, reconfigurable integrated circuits (ICs) could be used to provide a dedicated set of functions which are requested in certain driving situations. In case that one of these functions is not required any more, it could be dynamically replaced by another function which is requested and therefore needed by the driver. This process is called dynamic partial reconfiguration and was subject of research work in the past [90] [102]. Application Specific Integrated Circuits (ASICs) have hard-wired functions in silicon and offer very limited configuration possibilities only. For highly adaptive systems these kind of logic is of poor suitability, thus alternative solutions are of high interest. A chance to overcome the limits of ASICs are Field Programmable Gate Arrays (FPGAs). These special circuits are available from different vendors in various variations for covering a widespread field of possible applications and represent way more than just a chance to implement *glue logic* in a short amount of development time. However, their performance is poor when thinking about power consumption and resources efficiency [106]. In order to enhance and optimize FPGAs to be utilized in environments, where low energy consumption can not be neglected, special leakage current reducing measures must be added.

In addition to low power considerations, the question for secure design is a frequently discussed topic on various conferences [103] [104]. Security issues in classic office Information Technology (IT) are well known and therefore were handled mostly properly by appropriate usage of dedicated countermeasures, e.g., software updates. The awareness for dedicated countermeasures in, e.g., road vehicles, came up when special services like over-the-air (OTA) updates were introduced to meet the necessity of fast bug fixes in case of detected errors either in hardware or software or both at ECU level. Authentication and encryption are typically implemented in software, which is still the backbone for handling security incidents. However, by facing the progressive integration of more and more functions at chip level, it is not sufficient any more to use hard-wired functions in silicon for the purpose of accelerated encryption, but also to secure these chips against more advanced attacks, e.g., Differential Power Analysis (DPA), which is allocated to the category of side channel attacks. A careful analysis for optimization and hardening of sensitive ICs shall be done and used as starting point for a reasonable selection of appropriate design measures. Typical design constraints like performance, silicon area, low power and DPA robustness may contradict each other and highlight the fact of absence of an ultimate FPGA architecture which exhibits a complementary 100% satisfaction for all mentioned aspects. Waiving the importance of each design constraint heavily depends on the target application and should be prioritized case by case. Optimization always comes along with certain drawbacks to overcome during design iterations and requires special attention from first conceptual ideas down to practical implementation either at schematic entry or Register-Transfer-Level (RTL). Hence, a higher priority was put on the improvement of selected FPGA subparts for low power suitability and enhancement by security related features.

These topics, including all achieved benefits as well as collateral drawbacks, will be investigated and discussed in the following chapters of this thesis. As baseline architecture, the Xilinx Spartan

3A XC3S700A FPGA was chosen [4]. This is a low-end, commercial FPGA and therefore suitable for the cost-sensitive automotive industry. All following design measures and modifications are also suitable for the successor of this FPGA, the Xilinx Spartan 6 [24], which provides higher resources in terms of number of configurable blocks and a more advanced process technology ($45nm$ instead of $90nm$). It also provides a generic power-down mode that does not comprise special power-down options at a higher level of granularity. So the focus of this work was not put on improvements resulting from the application of a technology shrink, but on the benefits resulting from modifications and replacements done at circuit level.

In Chapter 2, FPGA basics are introduced and results of related academic research as well as selected commercial products are presented. In Chapter 3, the motiviation and basic considerations of low-power aspects are discussed. Chapter 4 explains basic aspects of cryptography and a selection of potential attacks. As FPGAs require memory cells for configuration, a selected number of memory cells and a new low-power memory cell is introduced in Chapter 5. A selection of existing flip-flop (FF) designs and a newly developed low-power derivate is discussed in Chapter 6, as this circuit type is required to buffer data if needed and an elemental part of each digital circuit. Before input data can be processed by the FPGA they must be fed into it, therefore a low-power optimized tristate buffer is presented in Chapter 7. In consequence, these components are integrated in Chapter 8. Last but not least, this thesis is concluded by a summary of all results in Chapter 9, followed by a statement about future work.

Chapter 2

Field Programmable Gate Array basics

2.1 Generic structure

After the invention of the first transistor, the idea came up to integrate entire circuits on a single piece of semiconductor. The first integrated circuit was introduced to the market in 1958 by Texas Instruments (TI) and was a phase-shift oscillator consisting of five integrated components. During the following years, integration of more and more transistors on the same silicon fabric went on and led to ICs containing basic logic gates to provide simple logic functions. In 1971, Intel presented the 4004, which was a microprocessor consisting of 2300 transistors and capable of processing more complex functions by executing software. The usage of this microprocessor was versatile as the software running on it could be used to implement customized functions. The continuous evolvement and improvement of microprocessors led to a wide range of types with high clock frequencies (up to $4.2GHz$), around 18 billion transistors and multicore architectures. In parallel to that, ASICs were developed to face special requirements of different applications in which microprocessors would not be the best choice. Once developed and manufactured, ASICs provide outstanding performance whilst still meeting constraints in terms of maximum power consumption and area footprint. However, both chip types do not have any option to change its internal design once manufactured in a foundry. All interconnects within the IC are hard-wired in silicon and can not be changed, which makes it impossible to fix bugs or to add desired modifications. This situation led to the perception that a new chip type was required, coming along with the ability to be reconfigurable in the post-production phase of its lifecycle, also called 'programmable in the field'.

Thus, in 1984, Xilinx released the first FPGA IC and kicked-off the development of a brand new product category. From that time on, different companies released a vast number of different FPGA types, which feature more or less capacity in terms of configurable resources and additional, dedicated blocks, e.g., multipliers, RAM or even entire microprocessor cores. A conceptual design of a FPGA is displayed in Figure 2.1 [71] [72].

5

Logic cell   Switch matrix

GPIO   Multiplier   RAM block

Figure 2.1: Conceptual design of a FPGA

To begin with, Figure 2.1 shows the internal, periodic structure of this IC, which is common for most FPGAs, despite their intended usage for either low-budget or high-end applications. The biggest portion of the chip area is used for logic cells which provide the ability to load the intended configuration into the FPGA. These logic cells must be connected to each other as it is likely that logic cell capabilities might need to be combined in order to implement more complex functions. For example, a simple logic function as shown in Figure 2.2 could be mapped into a logic cell. Basically, in order to implement this function, two logic gates (AND and OR gate) are required. The related truth table for this function is also shown in Figure 2.2. For configuration of a logic cell, the numbers of this truth table are transformed into logic values and loaded by a bitstream into the logic cell. As shown in Figure 2.3, the expected output values of Y are stored inside the configuration memory cells of a look-up table (LUT). At the same time, the inputs of the implemented logic function are used to drive the transistors of the multiplexer tree, which connects the output node Y to the configuration memory cells. So, in dependence of the relevant input values of A, $\overline{A}$, B, $\overline{B}$, C and $\overline{C}$, the correct, stored value of one of the eight memory cells will be transmitted to the output node. It should be pointed out that this is an example implementation, as the multiplexer tree could be also realized by using transmission gates instead of transistors. Figure 2.3 also implies the rising degree of complexity if more inputs need to be implemented:

6

the more inputs are used to implement a logic function, the more configuration memory cells and multiplexer tree stages will be required.

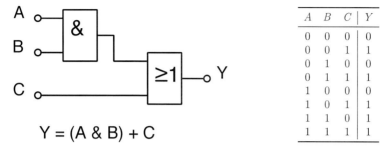

A	B	C	Y
0	0	0	0
0	0	1	1
0	1	0	0
0	1	1	1
1	0	0	0
1	0	1	1
1	1	0	1
1	1	1	1

$$Y = (A \, \& \, B) + C$$

Figure 2.2: Logic function Y implemented in logic gates and related truth table

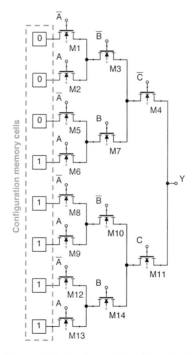

Figure 2.3: Configured LUT (simplified)

As the memory cells play a significant role inside a logic cell, these parts will be further evaluated and analyzed upon the possibility for further improvements in terms of power savings. Memory

7

cells are just one option among some alternatives how to realize programming of a FPGA. Table 2.1 displays a comparison of typical methods for configuration.

Technology	Volatile?	Re-Prog	Chip area	$R(\Omega)$	$C(\text{fF})$
SRAM	Yes	On-chip	Large	1-2k	10-22
Anti-fuse	No	No	Small	50-80	1-3
EPROM	No	Off-chip	Small	2-4k	10-22
EEPROM	No	On-chip	2 x EPROM	1-2k	10-22

Table 2.1: Characteristics of programming technologoies [72]

As described in Table 2.1, anti-fuse technology can be also used to program a FPGA by placing a layer of amorphous silicon between two metal layers. The amorphous silicon isolates both metal layers from each other, therefore a current flow between these layers is avoided, depicted in Figure 2.4.

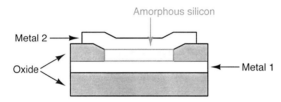

Figure 2.4: Anti-fuse Programming Technology (ViaLink) [72]

If a sufficiently high voltage (approximately $10V$) is applied on the anti-fuse, the amorphous silicon changes its state and provides a link between the metal layers. Hence, a resistance drop from $1G\Omega$ to 80Ω is created and the logic cell is (partly) programmed. Area occupation of this technology is comparably small and therefore the amount of parasitic capacitances as well. The downside here is that this step is irreversible and requires special transistors to create the required programming voltages, which results in higher power consumption. Anti-fuses require a special manufacturing process, which also adds more efforts and cost increase in comparison to standard memory cells. EPROM (Erasable Programmable Read-Only Memory) can be also utilized to program reconfigurable logic. Special floating-gate transistors provide the ability to trap electric charge on a floating gate, shown in Figure 2.5.

Thus, by trapping a charge under the floating gate while causing a large current flow through the transistor, the EPROM transistor remains permanently turned off. Re-programming can be performed by exposing the gate to ultraviolet light, so that the trapped electrons can drain through the gate oxide into the substrate. The functionality of an EPROM transistor inside a FPGA is also displayed in Figure 2.5: as long no charge is trapped on the floating gate and if a sufficiently high voltage is applied on the wordline, the transistor remains in an ON state and can be used

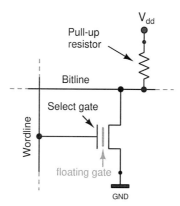

Figure 2.5: EPROM programming technology

to pull the bitline to a logic zero / *GND*. So in contrast to anti-fuses, this FPGA type can be re-configured, but this can not be done in-circuit and requires additional equipment. EEPROM (Electrically Erasable Programmable Read-Only Memory) goes back to the same operating mode, but provides in-circuit reconfigurability. Nevertheless, a drawback of EEPROM technology is the high area occupation and necessity for different voltage levels to be generated for reconfiguration. For the sake of reconfigurability, memory cells like SRAM (Static Random Access Memory) provide the most interesting set of features, as they usually have a low power consumption, provide the option for unlimited reconfiguration and do not require additional equipment for that purpose.

It is likely that functions to be implemented inside the FPGA will be way more complex than mentioned in Figure 2.2. For this reason, special switch matrices are added to the architecture and occupy a considerable portion of the chip fabric, as shown in Figure 2.1. An exemplary design of a switch matrix can be found in Figure 2.6 [72].

Based on the stored data inside the memory cells, the transistors of a switch matrix will be either turned on or turned off and therefore creating dedicated data paths to connect different logic cells with each other. This can be done between two logic cells next to each other or by long interconnect data lanes to link logic cells to each other which are spread over the whole chip. This interconnect logic may have more or less complexity, depending on the integration degree of logic resources. The detailed design of switch matrices is usually strictly protected by FPGA vendors, as this is considered intellectual property. Despite the fact that switch matrices are not in focus of this thesis, it is still worth to know that each improvement of memory cells in terms of power will also have a beneficial impact on these circuit parts. Interconnect metal layers are an integral part of each FPGA and can not be neglected when having parasitic capacitances and area footprint in mind. In consequence, a careful design of these metal layers must be done during the layout phase of a

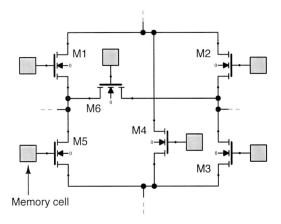

Figure 2.6: Exemplary switch matrix of a FPGA

newly developed FPGA. In dependance of the FPGA type, additional hard-wired blocks can also be part of the chip. Exemplary blocks are displayed by Figure 2.1, e.g., dedicated random-access-memory (RAM) and multiplier blocks. The configuration memory of the logic cells can be also used as RAM, however, dedicated RAM blocks provide in most cases faster read and write cycles. In addition to that, this solution also eliminates potential conflicts between configuration capacity and required memory storage. Multipliers, adders, etc. are very helpful if complex algorithms shall be designed and implemented on a FPGA. Instead of utilizing comparably slow logic cells for this goal, taken advantage of already existing, pre-defined blocks is certainly beneficial.

General purpose input/output (GPIO) blocks depict the interface to peripheral circuitry in a whole system. These parts are used to ensure reliable communication to, e.g., other FPGAs, memory blocks or any further kind of ICs. Protecting the sensitive FPGA logic from higher external voltages by shifting these voltages down to specified voltage levels to be further processed as well as shifting lower internal output voltages up to higher levels are essential tasks of GPIOs. Similar to the amount of interconnect layers, which proportionately increases to the overall complexity of a chip, the number of integrated GPIO increases as well as more input and output data must be transferred.

2.2 Typical applications

With the dawn of the first FPGAs, these chips started to cover more and more applications, which were typically handled by other IC types before, e.g., ASICs. Based on these facts, FPGAs exceeded the status of being glue logic and are frequently used in academic and commercial applications. A short extract of typical applications is shown below:

ASICs: The development of ASICs is time-consuming. High efforts must be taken into consideration as ASICs are intended to deliver best in class performance for the customers. Moreover, not detected faults in the design lead to high re-engineering costs and could be a reason why companies might encounter serious financial difficulties. In these cases, FPGAs are an interesting alternative as the time to market will be way shorter and any kind of fault which slipped through all verification steps can be easily corrected by reconfiguring the device. Of course, this way presumes that the chosen FPGA has sufficient logic resources, an adequate maximal clock frequency of optional dedicated hard-wired blocks and an acceptable price.

Prototyping: Fast prototyping can be achieved by taking advantage of existing FPGAs. As this chip is already available, there is no need to wait until any silicon will be manufactured by a foundry. Instead of this, the design of the desired IC can be synthesized by loading a Register-Transfer-Level (RTL) description of it into the FPGA. This leads to low cost of implementation and shorter development time. Existing prototypes can be also used as target hardware for firmware / software development at an early point of the overall project schedule.

Reconfiguration: Partial dynamic reconfiguration is of high interest if a high degree of flexibility is required. Even high-end FPGAs have limited capacities of logic resources, therefore (partial) reconfiguration during runtime of used resources can be a reasonable feature. A benefit of this ability is that operation does not need to be interrupted to load a new configuration into the FPGA and it is quite the contrary: the execution order of all necessary parts remains running. With that said, partial dynamic reconfiguration is a strong argument when costs and efficient power management take precedence over other aspects.

Sustainability: Older chips, which ran out of manufacturing and support of the vendor, might be still of high interest for various, critical applications (e.g. aerospace). Re-Manufacturing these chips in a more advanced process technology is usually no option due to expensive costs. FPGAs are an interesting alternative in this case as it is quite likely to have the RTL description of the discontinued IC. This RTL description could be then synthesized inside a FPGA in order to realize an alternative hardware solution rather than a software emulation of a generic microprocessor.

In summary, FPGAs offer many advantageous possibilities for different applications. They depict an important and meaningful supplement to the wide range of processing logic. A significant disadvantage of FPGAs is the fact that it is very difficult to fully utilize all internal resources, leading to an unused overhead of logic. Cost-efficiency is impacted by this circumstance, but above all, these unused parts consume energy without any reason. Despite the fact that software tools support energy-efficient synthesis of various designs, there is still room for improvement left. As already stated in Chapter 1, this and related topics were addressed during the research work for this thesis.

2.3 Existing designs

FPGAs are existing as commercial products in the market for many years already. A various number of vendors developed different variants of FPGAs for a wide range of applications. Some of these commercially available FPGAs already feature power saving mechanisms to make them more attractive for embedded applications in which a long battery lifetime is crucial. Keeping this in mind, for the purpose of this research it was important to highlight the FPGA types, which provide power dissipation reduction measures in order to check opportunities for even better implementations. Thus, selected FPGA types were chosen and are discussed in Section 2.3.2. In parallel, the work of other research groups that dealt with low-power aspects of FPGAs was also evaluated and reviewed. In fact, many results were published about reconfigurable logic and low-power applications, which are also shortly introduced in Section 2.3.1.

2.3.1 Related academic research

Over the past years, various research activities have been carried out to explore different possibilities for decreasing power dissipation of reconfigurable logic [39] [40]. The Xilinx Spartan 3A has already served as reference architecture in an earlier research project [42]. In that specific research, the chosen method was to combine different, common power optimizations, e.g., voltage scaling, power gating, low-leakage techniques for developing a new low-power FPGA. Different types of transistors with a different gate oxide thickness were selected to provide an efficient suppression of undesired leakage currents. If the selected manufacturing process supports these kinds of transistors, this method is very useful and should not be neglected when working on a design with decent static leakage current mitigation. Scaling down the supply voltage reduces power consumption and therefore a comparably low supply voltage was chosen [45]. This can be achieved either by implementing multiple, static supply voltages of different levels [52] or by dynamically scaling down or up the supply voltage on demand of the target application. In general, dynamic voltage scaling was discussed multiple times, as it is the right choice to scale down dissipated power in an efficient and fast way. The downside here is the amount of time spent on the implementation of a smart algorithm to trigger the power scaling in the right moment [47]. An alternative approach also takes advantage of the common power savings methods and combines these techniques with an operation of the FPGA in the subthreshold area [30]. This leads to ultra low-power operation of a FPGA, which might be of interest for internet of things (IoT) applications with very limited energy resources. Low-swing global interconnects, folded switch boxes and per path voltage scaling are also included. This approach delivers good results in terms of power savings, whilst slowing down maximum operation frequency of the chip. In another occasion, a different research group put their main focus on elaborating the minimum energy point of a new FPGA design. This energy point defines the lowest supply voltage provided to the FPGA, which still ensures reliable operation of the chip. For that purpose, a new FPGA design was presented based on a silicon-on-insulator (SOI) process [36]. Due to the low supply voltage, all memory cells used for configuration were replaced by latches for stable configuration data retention in subthreshold FPGAs. This is a promising solution for ultra low-power applications as long as circuit speed and area footprint play a negligible role. SOI manufacturing processes do not require special equipment for realization, however, these processes still require additional steps to be executed during the manufacturing procedure. FPGAs, which are based on this special technique are qualified to be used in applications with a higher degree of environmental stress, e.g., electromagnetic radiation in space applications. For that reason, additional research work was published, introducing a low-power and radiation-tolerant FPGA [34] based on full depletion SOI technology. In direct comparison to existing commercial solutions for applications with high electrical stress, this chip consumes less power but this advantage goes back to a modified manufacturing process and not on improvements at circuit level. On the search for new approaches for improving power dissipation of reconfigurable logic, a new LUT with non-volatile carbon nanotube based electromechanical systems (NEMS) was presented [54]. The main idea behind this design was to overcome the drawbacks allocated to a higher integration density (e.g., leakage currents and reduced reliability). Therefore, all configuration cells of this LUT were implemented in NEMS technology and merged with CMOS multiplexers as NEMS cells have almost zero leakage currents. Furthermore, as NEMS memory cells are of less complexity than their SRAM based counterparts, undesired parasitic capacitances were also reduced. Despite of the achievements in terms of power savings, this special manufacturing technology is not suitable for the mass market or cost-sensitive series production. In addition to that, conventional CMOS technology is also implemented on the chip to realize elementary circuit blocks like multiplexers, which leads to the necessity of both, NEMS and CMOS technology for the manufacturing process. An alternative approach to already existing CMOS and NEMS is to use magnetic RAM (MRAM) LUTs [37]. The core idea of this design is the introduction of a new spin-torque transfer magnetic RAM (STT-MRAM) cell for the implementation of a LUT. This cell is non-volatile even when the supply voltage is cut off, therefore re-programming is not necessary any more during the power-up phase. Furthermore, leakage power is minimized significantly. Based on that it can be stated that the STT-MRAM cell is certainly an interesting alternative for low-power applications, however, its reliability in operation does not achieve comparable results and still needs to be improved. A derivation of this technique is a non-volatile FPGA, which exploits the magnetic tunnel junction (MTJ) effect in combination with low-power measures [33]. Considerable power savings were achieved compared to classic SRAM-based results, although a comparison of the further aspects, e.g., maximum operating frequency and signal-to-noise ratio were not evaluated and also not compared.

Other research groups focussed on the design of more advanced non-volatile FPGA concepts by implementing additional logic utilized for energy harvesting within the chip. A hardware/software co-design approach is introduced and a checkpoint-strategy FPGA (CP-FPGA) is presented [38]. An integrated online-scheduler logic decides whether data stored within configuration logic needs to be stored into a backup area or not. This decision is taken by an analysis and so called checkpoint

13

identification of the executed application in advance, which subsequently leads to the necessity of special software. Despite this promising approach for low-power applications, the drawbacks of this solution is its dependance on pre-evaluation of the target applications by a software tool and the increased complexity in terms of additional scheduling logic, leading to a higher area footprint and higher programming voltages to load or erase data in non-volatile memory cells. Content addressable memory (CAM) is an interesting alternative to SRAM-based FPGAs. CAM can be used in applications where very fast access to stored data is required, therefore research was done to evaluate how CAM-based FPGAs could be realized [35]. Unfortunately, CAM comes along with shortcomings for low-power applications as additional logic is required to maintain short access times. Thus, improvements for reduction of consumed power were added to a CAM-based architecture. This led to positive results, however, these results did not depict a significant reduction of dissipated power.

A vast number of modern FPGAs use memory cells to store the desired configuration inside the chip [41]. These memory cells show leakage currents, which might turn into significant numbers when summed up. Hence, working on considerable leakage current suppression is a reasonable step. Different research works dealt with the improvement of look-up-tables (LUTs) [46] [48] [91]. At this point, the goal was to achieve a smaller power dissipation by evaluating new architectures for a power-saving LUT implementation. Adding multiple supply voltages to a LUT or entire architectural blocks was also in focus of research activities [52]. This is a reasonable step as the situation might occur where the highest computing performance is not demanded. Based on that, the circuit or gate can be driven by a lower supply voltage while extending battery lifetime due to less power dissipation. This technique was combined with a coarse-grain power gating of upper-level hierarchical blocks of the FPGA, after the pros and cons of fine-grain and coarse-grain power gating were evaluated [43] [51]. An elemental part of LUTs are multiplexers, which connect the configuration memory cells to the output nodes. Different optimization solutions were presented and compared in previous research work [31]. LUT topologies can be realized by various combinations of circuit structures to design 8:1 multiplexers. In dependance of the internal implementation, e.g., 8:1 multiplexer consisting of linked 2:1 multiplexers, power savings can be achieved. However, the related research results also show that these achievements depend on the input patterns. Optimizations of the multiplexer architecture are certainly reasonable as long as all related improvements are not subject to inputs.

Further research projects focused on the interconnect lanes and switching blocks inside a FPGA [39] [48]. As these interconnect layers and switching blocks occupy a large portion of the overall chip area [50], it is of high interest to design and implement them in a way that parasitic capacitances will be significantly reduced. Different ideas were explored and realized, e.g., efficient interconnect paths for minimizing the signal path delay between different logic cells [50]. This was also demonstrated by additional research projects [55], in which high-speed and low-power programmable interconnects were in focus of development. In this case, the combination of optimized interconnect lanes in terms of length and low-power techniques applied to buffers which drive the signals across the lanes, led to a reduction of dissipated power. Furthermore, an efficient design of switch matrices was also handled by research work [49]. FPGAs can be reconfigured, therefore a low-power design of switching logic might be beneficial if dynamic reconfiguration is frequently used. Each FPGA contains clocked logic inside its architecture, therefore it might be reasonable to implement clock-gating [44] [53]. If data does not need to be computed but to be maintained inside register cells, clock-gating is the first choice by switching off the clock signal and by stopping the operation of a circuit.

In summary, it can be stated that earlier research has showcased impressive results. Different topics and areas of a FPGA have been analyzed and potential solutions for improvement were elaborated. Nevertheless, a detailed analysis of these results and existing commercial FPGAs has revealed further opportunities for further improvements. The strategic approach of this thesis will be described after the introduction of selected commercial FPGAs from different vendors.

2.3.2 Commercial FPGAs

Market research has shown that companies like Altera / INTEL, XILINX, Microsemi, etc. provide commercial FPGAs with integrated low-power features. So there is a number of existing FPGA designs which have to be investigated upon their abilities to reduce power consumption.

Cyclone III: A direct competitor of the chosen baseline architecture is the Cyclone III FPGA from Altera / Intel. This product is allocated to the low-budget market and therefore also of interest for cost-sensitive applications. However, the Cyclone III design is manufactured in a $65nm$ TSMC process technology and has therefore a substantive advantage in comparison to the Spartan 3A competitor [5]. Due to that fact, a 1:1 comparison is not feasible but still worth to be taken into consideration. This FPGA's main power saving contributor is the manufacturing process itself, which offers a number of dedicated measures to extend battery lifetime: all-copper routing, low-k dielectric, multi-threshold transistors and variable gate-length transistors. Except the all-copper routing, which is used to increase the performance of the FPGA, all further measures can be seen as optimizations for power reduction. Altera goes one step further and combines these implementations with power mitigating algorithms in their Quartus II development software, which is used to configure a FPGA [6]. These algorithms feature a set of additional techniques to achieve further power savings, e.g., restructuring logic to reduce dynamic power, correctly selecting logic inputs to minimize capacitance on high-toggling nets, reducing area and wiring demand for core logic to minimize dynamic power in routing, modifying placement to reduce clocking power, etc. In dependence of the design which should be synthesized on the chip, the development software applies an appropriate selection of the feature set.

15

Cyclone V: This representative of the Cyclone family goes beyond the capabilities of the FPGA described before. Different variants (E, GX, GT, SE, SX, ST) of the Cyclone V FPGA are available for covering a wide range of thinkable applications [7] and some of them provide a high grade of complexity. For example, some Cyclone V types offer dedicated DSPs (digital signal processor) and CPU (central processing unit) cores within the FPGA fabric. Thus it should be made clear that these chips are rather allocated to SoCs (system on chip) than to low-end FPGAs. The aim here was to integrate a high number of different components as hard IP (intellectual property) blocks such as memory controllers with optional ECC (error correction code), PCI (peripheral component interconnect) express support, various transceivers, DSP blocks and a dual-core ARM Cortex-A9 processor (running at a maximum frequency of $925MHz$ [11]). By adding more logic as hard IP blocks on the die, power reduction can be achieved as these blocks are designed in an optimized way instead of using reconfigurable resources which is certainly not that efficient. Cyclone V chips are manufactured by an advanced $28nm$ process technology provided by TSMC [8] [9]. As expected, this FPGA family contains all power saving mechanisms which have been already used in previous generations, but takes it main advantage from the advanced manufacturing process, which allows better battery lifetime than provided by earlier products. This is supported by embedded hard IP blocks within the die and the ability for dynamic partial reconfiguration. The latter technique allows the reuse of blocks that change during system operation, thus giving the opportunity to re-configure logic resources by a new function, which have been occupied by another, unused function until that time. Hence, dynamic partial reconfiguration enables related FPGAs to be used in applications, which requested usually FPGAs with a higher density of logic resources.

Stratix 10: A further representative of the Altera / Intel FPGA products are the Stratix series. These reconfigurable chips are high-performance FPGAs, which easily outperform the Cyclone series due to their advantages in technology and design. This product provides a quad-core ARM Cortex-A53 processor, a memory management unit (MMU), more on-chip memory ($256KB$ instead of $64KB$) in the Cyclone V series and an ECC unit [11]. The maximum achievable frequency of the integrated CPU is $1.5GHz$ and therefore also remarkably higher than the maximum frequency provided by the Cyclone V [11]. Similar to the Cyclone V family, Stratix 10 FPGA are also SoCs but manufactured in a more advanced $14nm$ Intel Tristate-Gate process. This attribute adds further power savings when compared to any other design manufactured in an older process technology as smaller structures lead to less parasitic capacitances. So at this point, the Stratix 10 FPGA takes advantage of the latest advancements in process technology. Extreme low power variants are available by setting the internal supply voltage to a fixed voltage of $0.8V$ [10] by accepting penalties in performance. In case that more computing power is required, dynamic scaling of the supply voltage in other variants of this chip. Furthermore, power gating is implemented to switch of different components of the SoC, if their function is not demanded any more. For example, the internal DSP and memory blocks can be switched off by using power gating and therefore preventing undesirable leakage currents inside the chip. This method is very efficient and should be used whenever its usage is meaningful and still consistent with the intended function of the implemented design. This method is supplemented by clock gating for dynamic power reduction. Switching off the clock signal of a logic block leads to a hold of the operation and therefore to less power consumption. As complex FPGAs require adding complex and fine-grain clock trees, this method is useful to set whole blocks of the chip to a hold status.

PolarFire: Microsemi has developed a set of FPGAs to cover different customer demands. At this point, the PolarFire FPGA family is of interest as it was designed for low-budget applications where low power consumption is of high interest. This chip is manufactured in a special UMC $28nm$ non-volatile process technology which adheres floating gate non-volatile transistors instead of memory cells to implement the intended configuration in the logic [12]. Non-volatile transistors have the advantage to be used as electron trap by adding special oxide layers to avoid the electric charge to leak out of the device. Once these transistors are programmed and once the the electric charge is trapped inside these devices, leakage goes down and contributes to the reduction of dissipated power. However, the downside of this solution is that high voltages are required to program and to erase each of these special transistors as a remarkably high oxide thickness must be bridged by the programming / erasing voltage. These high voltages require integration of charge pumps to generate them, which comes at cost of power and area consumption. Despite the fact that this technology leads to power savings, all additional steps to integrate further oxide layers raise the manufacturing costs and have to be considered in cost-sensitive projects. A special feature of this certain PolarFire FPGA variants (all S products) provide security-related features, such as DPA bitstream protection, tamper detectors, secure non-volatile memory, a crypto coprocessor, etc. This integrated coprocessor can be used to, e.g., accelerate encryption by using well known crypto algorithms and therefore to provide an interesting set of security-related functions [13]. Furthermore, hard IP blocks are also an integral part of these FPGAs, e.g. the math block. This hard IP can be used to speed up the computation of mathematical operations and supports implementing finite impulse response (FIR) filters, which leads to extra power savings [12]. The supply voltage ranges from $1V$ to $1.05V$ if a higher performance is requested, but does not drop below $1V$ for achieving an ultra low power mode.

iCE40 (Ultra/LM/LP): Lattice Semiconductor designed a series of FPGAs for use in portable and battery-powered devices. The main task of these devices is to support the main processor in a system by executing operations which are not time critical and can be handled by the FPGA while setting the main processor to a low-power mode or idle state. This leads to the situation of having a power-optimized FPGA and a main processor in a power saving mode [14]. Different variants of the iCE design have been released, an extract of these are

17

the following products: iCE40 Ultra, UltraLite, UltraPlus as well as iCE40 LP (low-power) and LM (low-power with integrated). These types have been selected because they depict the low-power derivates of the FPGA line-up in this company. All of them are produced in a $40nm$ TSMC process, which provides considerable out of the box power savings when compared to older process nodes [15] [16]. The iCE Ultra series were designed to be used in wearable technologies, so the strategy to battery lifetime extension is to add hard IP cores to the FPGA fabric. Similar to the described approach of the Cyclone V hard IP integration, the core strategy for saving battery power is to add custom hard IP blocks for dedicated functions, e.g., inter-integrated circuit (I^2C) bus, serial peripheral interface (SPI) and DSP blocks [17]. The iCE40 LP and LM variants do not contain any hard IP blocks and therefore take advantage of the $40nm$ CMOS low-power process technology only. In contrast to the Ultra variants, these FPGA types provide more configurable resources, more embedded memory cells and a higher number of I/O (input/output) pins. However, the lack of dedicated hard IP cores leads to the necessity of implementing each required block as so called *soft core*. This solution usually comes along with a higher power consumption as their design is based on usage of the logic resources and not on full-custom design and layout. On the other hand, the abstinence of hard IP blocks lead to a lower price and therefore might raise the interest of cost-sensitive customers.

ArcticPro eFPGA: QuickLogic offers the ArcticPro eFPGA, which is a SoC with an embedded FPGA core on the die. The company has chosen an alternative way to develop this chip by adding reconfigurable logic to a SoC instead of building a SoC around the FPGA fabric. Hard IP blocks like an ARM Cortex-M4 CPU including a coprocessor, memory areas, I^2C, SPI, a sensor manager, etc. represent fundamental components of a SoC, which also do not constrain any resources of the reconfigurable logic. The idea behind this approach is to allow designers to configure / customize functions in the post-production phase without expensive and time-consuming redesign [18]. ArcticPro eFPGAs are promoted as ultra-low power solutions, which take advantage of different manufacturing processes in $65nm$, $40nm$ and also in $22nm$ provided by SMIC. Power savings of this chip will get better by moving to smaller process nodes and are the main argument for decreasing overall power consumption. This FPGA is a versatile solution for a wide range of applications. It is not a pure low-end FPGA as it implies complex hard IP blocks to realize whole systems on this chip. On the other hand, it is not a pure high-end FPGA for applications where high computing performance is required.

Spartan 6: The Spartan 6 FPGA was developed by Xilinx as a low-cost and low-power solution for fierce applications in which energy and budget play an important role. This design is a successor of the original Spartan 3 chip and therefore comprises improvements for the sake of extending battery lifetime. Similar to the competing FPGAs described until this point, the most significant contributor to less energy consumption was the process technology change from $90nm$ down to $45nm$. This process shrink comprises almost the same low-power related benefits which have been already introduced earlier this chapter. Furthermore, meaningful architectural elements of the Virtex 5 design were entailed to the Spartan 6 series, e.g., the 6-input look-up tables instead of the 4-input look-up table of the Spartan 3 series [19]. The supply voltage was decreased from the previous $1.2V$ down to $1V$ and certain hard IP blocks were also added, e.g., a memory controller for quick and easy design of memory interfaces, transceivers, PCI express for faster connectivity (Spartan 6 LXT) and dedicated multipliers, accumulators, pre-adders/subtractors and wide counters for efficient implementation of DSP algorithms [20]. Xilinx decided to use a mixture of low threshold and high threshold voltage transistors across the entire chip. The reason for this decision was to use slower, high-leakage transistors for less time-critical areas like configuration blocks and to use faster, low-leakage in blocks where execution time matters. An option to shutdown the phased-locked loops (PLLs) is also integrated in case that clock gating was applied before and the PLLs do not need to be powered any longer [21]. As a last point, it should be also mentioned here that the Vivado and ISE design suites also support dynamic power reduction measures during synthesis of a design.

Spartan 7: The Spartan 7 series FPGAs are improved variants of their predecessors, the Spartan 6 FPGAs. Xilinx moved from the previous $45nm$ manufacturing node to a more advanced $28nm$ high performance, low-power technology (HPL) node provided by TSMC [23]. However, architectural aspects of the 7 series were also adapted and added to the Spartan 7 design, leading to additional dynamic power savings of approximately 50%. So it can be summarized that static power reductions are only from process enhancements and capacitance reductions (geometry shrink and low-K dielectric). Taking a closer look on dynamic power reductions, it can be stated that architectural enhancements like clock gating, 6-input look-up tables, hard IP blocks, and system-level power management features are used to achieve better results [23].

Artix 7: Xilinx has recognized market demand for FPGAs with capabilities to save power during runtime and to deliver acceptable performance as well. In order to close the gap between the Spartan Series and the Virtex Series, the Artix Family was introduced. All applications, which require a SoC platform and show aggressive cost constraints are often coupled with low power requirements, typically due to the small form factor and high-volume nature of these end-applications and respective markets. The Artix 7 series consists eight variants, which differ in terms of complexity like number of available logic cells, number of transceivers, etc. A substantial basis for these FPGAs is TSMCs HPL process technology, which ensures out of the box power savings for each manufactured design and which features high-k metal gate (HKMG) transistors [25]. Except the Spartan and Zynq series, the Artix, Kintex and Virtex are manufactured with the same $28nm$ process node and share many architectural elements, which allow scalability among these series. This results in easy upward and downward

19

migration across different programmable devices and families [27]. For example, the same design which was implemented in a Kintex 7 or Virtex 7 FPGA can be also implemented in a Artix 7 chip if slower operation can be accepted in the respective application, while gaining better results in power dissipation. This process node also allows more voltage headroom as it supports supply voltage scaling by a low voltage mode (down to $0.9V$) and a high voltage mode at $1V$. Adaptive voltage scaling (AVS) provides the option to control and to scale the supply voltage by the customer and thus to add an extra amount of power savings or performance if required. Several measures are added to the overall chip design to mitigate static power dissipation, such as power gating, a set of common integrated hard IPs, stacked silicon interconnect (SSI) technology, and partial reconfiguration [26] [27]. In addition to that, clock gating is available to reduce dynamic power and is therefore supported by the Vivado and ISE design tools [27].

Virtex 7: Xilinx' Virtex 7 FPGAs are high-end solutions for reconfigurable logic and provide highest system performance, capacity while still providing the same power saving features of the 7 series family [27] [29]. Nevertheless, the focus of these devices are not primarily applications with limited energy resources but applications with a high demand of fast computing abilities. Different Virtex 7 variants have been developed and they differ in number of available logic cells, internal memory cells and DSP slices (Spartan 7 XC7S100 provides 102,400 logic cells whilst Virtex 7 XC7V2000T has 1,954,560 logic cells built-in [29]). DSP slices can be used to implement desired DSP functions by providing dedicated multipliers and accumulators to enhance efficiency and speed. These aspects allow designers to use Virtex 7 FPGAs for ASIC prototyping to speed up the development process and to decrease the time to market.

In summary, it can be stated that FPGAs have entered the low-power market by featuring different mechanisms for extended battery lifetime. The most significant achievements in power savings rely on advancements in manufacturing technology as these designs take advantage of low-power benefits which come along with process shrinks. Second, integration of hard IP blocks for dedicated measures are listed as an efficient way to save both, battery power and logic resources as these blocks do not have to be developed by the customers but are ready to be used right after powering on the FPGA. Gating techniques, such as power gating and clock gating are mentioned as used design measures inside the fabric to keep static and dynamic power dissipation under control. At architectural level, the shift from 4-input to 6-input LUTs is described as meaningful step forward to minimize signal propagation delay and therefore to raise system through-put [20]. A simplified implementation is depicted in Figure 2.7.

Figure 2.7 shows an example implementation of both solutions and highlights the differences in complexity. Despite the fact of more required multiplexers and registers, a 6-input LUT might be more suitable to implement complex functions within one logic cell. In contrast to that, it is likely

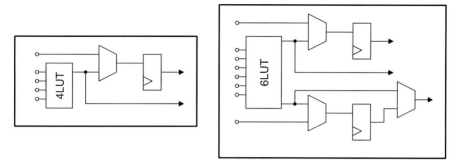

Figure 2.7: Comparison of 4-input LUT and 6-input LUT [20]

that the same implementation will require more than one logic cell with a 4-input LUT.

By referring back to the Spartan 3A baseline design of this thesis, the main focus was to decrease power dissipation without taking advantage of a more advanced manufactured process. Hence, the same process technology (TSMC 90nm) was used for all following research and development work. The chosen approach here was to modify and to improve selected parts of a FPGA by applying power saving and leakage reduction measures or by implementing completely new power-optimized designs in these parts. Some of the techniques introduced before were also used during the research and combined with new re-designed circuits, which provide a lower out of the box power dissipation. Integration of hard IP blocks was relinquished as the design should be as simple and cost-efficient as possible. This strategy also led to the decision to use 4-input LUTs instead of 6-input LUTs. Any given design can be implemented with less silicon area using a 4-input LUT than using a 6-input LUT [12]. Each LUT needs to be configured by memory cells, storing configuration bits to load a function into the logic cell. Whilst a 4-input LUT requires 16 memory bits, a 6-input LUT requires 64 memory bits which is 4 times as much memory but can accommodate only about 1.6 times as much logic as the smaller 4-input LUT [12]. Beyond this, basic gates like AND, NAND, NOR, etc. functions are also part of logic blocks inside a FPGA and should not be neglected when focusing on overall power optimization. Hence, these elements were also analyzed and modified. A fine-grain approach for power- and clock-gating was added to allow full control down to particular cells with the goal to either power-down unused circuits or to stop their operation while still ensuring data retention.

2.4 New approach in this thesis

In contrast to previous academic research work, during this study, highest priority is put on modifications or complete redesigns at gate-level. It was not intended to rely on advanced techniques like MRAMs, NEMS or carbon nanotubes; the chosen approach in this work was quite the contrary: all power savings should be achieved by implementing power reduction measures together on redesigned or newly developed basic gates / cells. It is simple to achieve lower static and dynamic power consumption by synthesizing a circuit design with a modern process technology. So, before referring back to this method, modifications were done at gate level to embed low-power optimization from schematic entry on. Also, it was not intended to operate the cells at the cutting edge of supply voltage. There are no doubts that decreasing the supply voltage remarkably reduces consumed power by accepting astable working conditions at the same time. Therefore, a reasonable scaling of external and internal voltage nodes was applied. A new memory cell design was developed and modified with common low-leakage techniques. The combination of both, redesign at schematic level and added power reduction measures, led to the introduction of a new memory cell which was not presented before. A new low-power LUT was implemented based on these newly developed memory cells, which also led to the necessity of low-power adaptions of basic gates (as described before). The same procedure was applied to design a new data flip-flop (D-FF), which is frequently used inside a FPGA. The basic idea was to select a different logic style and to add power reduction measures for achieving both: low power consumption and a considerable operating frequency. The usage of a differential and dynamic logic style, added a security-related feature to this D-FF, which is advantageous for FPGAs used in security-related applications. Also, this design uses energy-harvesting to contribute to battery lifetime extension. This combination was not handled by earlier publications. A further achievement was the design of three tristate buffers with different pros and cons. Again, an existing legacy design was analyzed and improved by applying redesign steps, modifications at circuit level and further low-power techniques. During the next step, the newly developed memory, D-FF and tristate buffer were very useful for designing a new low-power GPIO buffer. All integrated, low-power components led to a new GPIO buffer, which was also not introduced by other research groups before. The same statement is also valid for the realization of a logic cell, which is based on all previously developed designs and therefore also depicts a new implementation.

In summary, redesigning integral FPGA parts for achieving substantial, intrinsic power savings by design and adding static and dynamic power reduction optimizations, distincts this research work from previous applications. Whenever possible, other circuit characteristics were also considered and improved.

Chapter 3

Aspects of low power applications

3.1 Motivation

Back in the days when power dissipation was not the biggest concern in circuit design, two other characteristics had a higher priority in the designers' minds: performance in terms of the IC's speed and consumed area on the silicon wafer due to a preferably high yield. Thus, power dissipation was not more than an afterthought for many years. However, the upcoming mobile devices like smartphones, smartwatches and, of course, higher demand of computing power in vehicles led to a change in prioritization of design aspects for ICs [3]. Designers, being involved in engineering low-power chips, have the option to choose power-saving measures allocated to different layers, shown in Figure 3.1 [50].

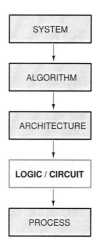

Figure 3.1: Power optimization hierarchy

System: According to Figure 3.1, certain design options can be applied at system level, to prevent a chip from draining more current out of the battery than is needed. One way for realizing power optimization at this level is to enforce integration of different ICs / functional blocks to one DIE. The history of microprocessor development has followed this approach in a very successful way: the Intel 80386 processor consists of two separate chips, namely the main processor (i386) and the co-processor (i387), which represents the floating-point unit (FPU). The next descendant of that processor, the Intel 80486 exhibits an integrated FPU. This step was very beneficial as a high integration comes along with less parasitic capacitances, less circuitry overhead (what also contributes to mitigation of parasitic capacitances), less efforts left for manufacturing and lower costs. Over the time, more and more former dedicated ICs were integrated, e.g., 2nd-level cache was added on the DIE of microprocessors, the graphic processor unit (GPU), the digital clock management (DCM), the memory management unit (MMU) and so on. This is shown in Figure 3.2 by way of example.

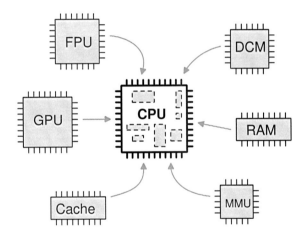

Figure 3.2: Rising degree of integrated components

For a long time, progress in operating speed was achieved by a continuous boost of the system clock. However, the physical characteristics of silicon limit the maximum operating frequency at approximately $4.2GHz$, so other ways to increase operating speed must be found. Apart from that, a high clock frequency comes along with a higher temperature of the chip which requires appropriate cooling. Parallelization and keeping the system clock frequency comparably low are ways to overcome these limitations at system level.

Algorithm: The next option to work on the low-power aspects of a circuit is to take a closer look on the algorithm, which is implemented in the design. Optimizing the algorithm in terms of power means to cut down the number of executed operations and therefore also the hardware resources in use [92] [93]. These activities have to be carried out at a very early point of the development process, as it might be difficult to modify algorithms at a later point of time. Also, coding data in a way that minimizes the circuit switching activity is a method, which should be also allocated to algorithmic improvements.

Architecture: Going further along the top-down flow of Figure 3.1, the next hierarchy level is the architecture of a design. An efficient power management is the basis for implementing a combination of coarse-grain and fine-grain gating principles. Coarse-grain power management is realized by the ability to shut down all blocks, which are currently unused and therefore should be completely powered down in order to prevent flow of leakage currents through these blocks. In case that data storage is required, clock gating can be a less eradicative but also efficient method to prolongate battery capacity. A further extension of coarse-grain power gating is fine-grain power gating, which will be introduced in the following chapter, as it is rather a method which is applied at logic / sub-circuit level. The basic principle of coarse-grain and fine-grain power gating is depicted in Figure 3.3

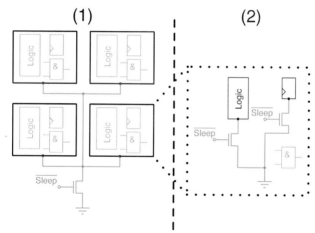

Figure 3.3: Simplified illustration of coarse-grain gating (1) and fine-grain gating (2)

Logic / Circuit: In case that coarse-grain gating is not sufficient for the designers' goals, a more detailed level of gating can be applied: fine-grain gating. Shutting down sole blocks of a certain design might be reasonable if stronger leakage current suppression is demanded or if there are sub-blocks inside the design, which are not necessary for further operation, e.g., registers for input and output data. Another application case might come up if a total shutdown of the supply voltage is not required, but a reduced supply voltage is used instead. This method is known as voltage scaling and can be used during runtime as long as a stable operation of the circuit is ensured by a lower V_{dd}. Full custom design of a circuit provides the chance to minimize parasitic capacitances in a better way than using standard cells from a technology library for the same purpose. Working on a full custom design also allows to choose between transistors with different threshold voltages, which support to suppress undesirable leakage current flows.

Process: Advancements in process technology made design shrinks feasible. Moving from an older process technology to the latest manufacturing process allows to reach out for a wide range of advantages, e.g., higher integration density, availability of multi oxide thickness and multi threshold voltage transistors, a possible reduction of the transistor count, reduced parasitic capacitances by smaller geometries of the layouts and many more. Process technology is getting improved continuously, so that it is just a matter of time and available budget to get the latest benefits of process evolvement.

It might be difficult to apply power reduction measures to each hierarchy layer of Figure 3.1. However, this was not the intended goal of this thesis, as the same figure also highlights the main focus of this research work: low-power improvements for different blocks of reconfigurable logic at circuit level. How this was applied to selected components of a FPGA will be further explored and described in the following chapters. Every circuit depends on the availability of a supply voltage and the current, which is taken from the battery source and flows through a circuit, ensuring its proper operation. All chips dissipate a certain amount of power during runtime, so that two different types of power dissipation can be identified:

- Static power

- Dynamic power

Having the Xilinx Spartan 3A FPGA in focus, the breakdown of static power dissipation is displayed in Figure 3.4 [91].

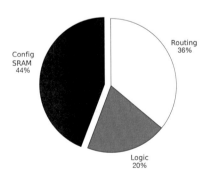

Figure 3.4: Breakdown of static power dissipation

26

First of all, Figure 3.4 shows the partitioning into three different types of circuit blocks: the pure logic, configuration memory cells and the routing, which is the interconnect portion of the FPGA. So according to this figure, a percentage value of 44% of the overall static power dissipation is allocated to the memory cells, which configure the basic functions of reconfigurable logic. This result clearly indicates that leakage current minimizations are worth to be applied on these cells, as considerable improvements can be expected here. The pure logic cells consume less power, as they remain in standby mode and do not charge or discharge load capacitances. Nevertheless, 20% of the overall static dissipated power is still a number, which can not be neglected and should also be improved. Last but not least, the remaining 36% of dissipated power is caused by the routing cells. There is no doubt that interconnect logic should also be carefully analyzed in order to see how the 36% can be decreased to achieve an acceptable level.

As each circuit will also be executing a certain function, dynamic power dissipation will come into play. Based on that, the related distribution to different blocks is of high interest as shown in Figure 3.5 [91].

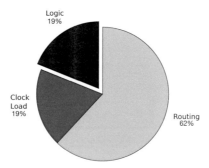

Figure 3.5: Breakdown of dynamic power dissipation

In contrast to Figure 3.4, configuration memory cells do not play a role in Figure 3.5. The reason for this fact is that these parts do not affect dynamic power dissipation during runtime, as they are configured in the power-up phase and then not touched again, unless reconfiguration will be requested. However, configuration cells do not contribute significantly to overall dynamic power dissipation. Instead of this, the core logic is way more important. According to Figure 3.5, core logic takes 19% of the overall power dissipation and must be considered when talking about power-efficient use cases. Here, core logic is active while the input values are processed and therefore also calculated to provide the output data. Based on this awareness, logic shall be optimized for both, static and dynamic power dissipation. Another portion of the total power dissipation is generated by the clock load and takes 19%. Clock load power dissipation is difficult to mitigate as it arises at the time when clocked circuit parts charge and discharge load capacitances. This can be handled by careful layout of the schematics, but requires high effort. For timing critical applications higher attention should be put to this aspect. This most significant portion (62%) shown in the circle diagram of Figure 3.5 is allocated to interconnect circuitry. In consequence, low-power design of a FPGA's switch matrices is obviously a reasonable task and will be handled in future work, but is out of scope here. The reason for this decision is that FPGA vendors protect the internal design of the interconnect circuitry as it is their intellectual property. Therefore it is challenging to do an in-depth analysis of these parts. In summary, it can be stated that optimization of the static power losses comes first and further optimization of dynamic power comes second. This approach was followed throughout the entire research work.

3.2 Static power dissipation

Minimizing the total amount of power consumption leads to the estimation of all contributing components, like described before in Section 3.1. Regardless of static and dynamic power and before going deeper into discussing the dependencies and the impact of it, it shall be made clear how to distinct between the *maximum* and the *average* power dissipation. The first one is related to the maximum peak current, whereas the latter one is the average dissipated power during a certain simulation / runtime $t_{average}$. As already stated, the most important motivation behind this work is to apply power reduction measures for the sake of battery lifetime extension. Thus, the amount of average power is more important for the battery lifetime than the peak power. In contrast to this, peak currents impact the supply voltage only due to to the power line resistance, which may cause heating of the chip and degradation of performance. Based in the example of a simple CMOS inverter, shown in Figure 3.6, Equation 3.1 depicts both summands of static power [116].

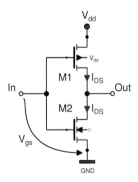

Figure 3.6: CMOS Inverter with static I_{DS} current

$$P_s = P_{s1} + P_{s2} \tag{3.1}$$

So, Equation 3.1 shows P_{s1}, which is the outcome of leakage currents and P_{s2}, which is the result of $I_{V_{dd}}$ in case that the input voltage V_{GS} is higher than 0. Leakage currents arise through MOS junction leakage currents, flowing through parasitic diodes inside of a CMOS inverter. The current I_d in both of these diodes can be described by Equation 3.2.

$$I_d = I_s(e^{\frac{qV_d}{nkT}} - 1) \tag{3.2}$$

The coefficient of a diode is given by n, the saturation current by I_s and the applied voltage is given by V_d. Furthermore, k is the Boltzmann constant, q the elementary charge and T the absolute temperature of the p-n junction. It can be also seen by Equation 3.2 that I_d is dependent of the temperature and increases by applying a higher temperature. Therefore a decent cooling of semiconductors is usually a reasonable measure. Based in that, Equation 3.3 represents the calculation of P_{s1}.

$$P_{s1} = \sum_{i=0} I_{di} V_{dd} \tag{3.3}$$

In order to keep P_{s1} as low as possible, lowering V_{dd} is certainly a good approach as this has a direct throughput on undesired power consumption. Keeping the temperature down will also be a supportive design measure, as well as keeping $V_{sb} > 0$ to prevent leakage currents of becoming too high. The next summand of Equation 3.1 depends on the input voltage V_{in}, shown in Equation 3.4.

$$I_{DS} = I_0 \frac{W_{eff}}{W_0} 10^{\frac{(V_{in} - V_{th})}{s}} \tag{3.4}$$

Here, I_0 is the current when V_{gs} equals V_{th}, W_0 the width of a transistor without considering etching trenches and s is the subthreshold swing paramter, which is the gate voltage swing required to reduce the drain current by one decade and has a theoretical minimum limit of $60mV$ per decade. It may happen that small V_{gs} voltages will be accidentally applied to the input node of a CMOS inverter. If these input voltages are smaller than the threshold voltage V_{th} of a transistor, then Equation 3.4 can be used to calculate I_{DS}. It shall be noted here that the effective width W_{eff} of a MOSFET transistor has a high influence on the result of I_{DS}. Thus, keeping W_{eff} as small as possible will also help to push I_{DS} further down, but it will also slow down the maximum operating frequency. So making the assumption that the input voltage is $0 < V_{in} < V_{th}$, the related power dissipation P_{s2} can be outlined by Equation 3.5.

$$P_{s2} = I_{DS_{mean}} V_{dd} \tag{3.5}$$

In Equation 3.5 I_{DS} is the mean value of the current flowing through both $M1$ and $M2$ of Figure 3.6. As stated before, maintaining the transistor width supports the reduction of P_{s2} as well as choosing transistors with a high threshold voltage V_{th}. During the research work, a careful selection of these described measures was done to achieve a considerable improvement of static power consumption. This will be further explained in the next chapters.

3.3 Dynamic power dissipation

Every chip will be connected to another analog or logic block and therefore also driving the input stage of this block. This also means that this input stage of the following block can be seen as a capacitive load, which will be either charged or discharged. An exemplary situation is illustrated in Figure 3.7 [116].

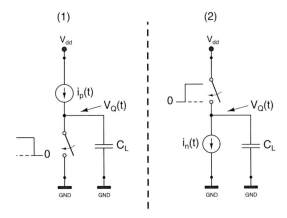

Figure 3.7: Equivalent circuit for power calculation

As shown in Figure 3.7, either current i_p will charge the output load C_L or current i_n will discharge C_L to GND. These processes will subsequently lead to power dissipation, which will also depend on the switching frequency f, shown in Equation 3.6 [109].

$$P_d = \frac{1}{T} \int_0^T i_q(t) v_q(t) dt \tag{3.6}$$

30

Depending on the situation (charging / discharging) the output current is given either by

$$i_q = i_p = C_L \frac{dv_q}{dt} \tag{3.7}$$

or by

$$i_q = i_n = -C_L \frac{dv_q}{dt} \tag{3.8}$$

which leads to the following results

$$P_d = \frac{1}{T} \left[\int_0^V C_L v_q dv_q - \int_V^0 C_L v_q dv_q \right] \tag{3.9}$$

$$P_d = \frac{C_L V_{dd}^2}{T} = C_L V_{dd}^2 f \tag{3.10}$$

Equation 3.10 is usually extended by an activity factor α for complex designs. So it can be noticed here that power dissipation strongly depends on the supply voltage V_{dd} and the operating frequency f. In case that reduction of dynamic power consumption is required, these factors and the load capacitance C_L shall be taken into consideration. Reduction of C_L might be difficult as this might not be in the designers' hands. However, tweaking the supply voltage is more likely, because every decrease of the internal supply voltage can be compensated by appropriate level restorers. An interesting fact here is that dynamic power dissipation can be mitigated not only by supply voltage gating or supply voltage scaling, but also by slowing down the operating frequency or by setting it to $0Hz$. The above described measures were in some cases implemented in the newly developed designs as will be described in the following chapters. Those measures were used only when their application was considered meaningful to the overall design.

3.4 Leakage current suppression

In Section 3.2 and 3.3 aspects of power dissipation and its components were introduced. Going a little bit more in to detail means to analyze which factors directly impact leakage currents in the subthreshold region of a transistor and to see, what kind of countermeasures may be to implemented. So in total, three major components of leakage currents can be identified for a Metal-Oxide-Semiconductor (MOS) transistor of gate lengths in nanometer scales [97]:

- Subthreshold leakage
- Direct tunneling gate leakage

- Reverse biased p-n BTBT leakage

Whilst the band-to-band tunneling (BTBT) leakage currents can be neglected for devices exceeding 50nm gate lengths, subthreshold and direct tunneling gate leakage currents come into consideration for our design. Tunneling electrons through gates oxides can be countermeasured by carefully setting an adequate oxide thickness of each transistor. This dependency can be seen in Equation 3.11:

$$J_{DT} \propto A(\frac{V_{ox}}{T_{ox}})^2 \tag{3.11}$$

$$A = \mu_o C_{ox} \frac{W}{L_{eff}} (\frac{kT}{q})^2 e^{1.8}$$

The gate capacitance (including the oxide) is represented by C_{ox} and μ_0 is the magnetic constant. By increasing the oxide thickness T_{ox} and keeping the voltage over the oxide (V_{ox}) at the same level, the direct tunneling current density J_{DT} can be efficiently lowered to a minimum stage [1]. Increasing the gate length L_{eff} would have a similar effect, but will lead to higher effort in the manufacturing process due to a change in one of the basic technology parameters like the gate length of a transistor. Therefore, this option should be avoided. However, the usage of multi-oxide thicknesses is a technology dependent parameter and requires awareness for the selection of a suitable multi-oxide technology.

Subthreshold currents I_{sub} can be expressed by the following Equation 3.12 [48]:

$$I_{sub} \propto \frac{W}{L_{eff}} e^{(V_{gs} - V_{th0} - \gamma V_{sb} + \eta V_{ds})/nV_t h} (1 - e^{-\frac{V_{DS}}{V_{th}}}) \tag{3.12}$$

Equation (3.12) shows the parameters which contribute to the overall weak-inversion current, flowing below the threshold voltage V_{th} of each MOS transistor in the circuit. In this equation, γ and η are scaling factors for the respective voltages Several leakage reduction measures can be applied by utilizing these parameters to design a low leakage circuit:

- W: setting the width of a transistor as small as possible leads to a higher resistance of it and therefore to smaller leakage currents

- V_{gs}: Gate biasing is done by applying a V_{gs} voltage lower than GND, which turns the transistor deeply off

- V_{sb}: Body biasing by tweaking the body voltage of a turned off transistor

- V_{ds}: Drain-source voltage, which can be decreased by applying stacking in the pull-down network of a logic gate

- V_{dd}: Lowering the supply voltage mitigates or even completely removes the DIBL (drain-induced barrier lowering) effect, represented by η in Equation 3.12

In general, we can distinguish between two classes of leakage reduction techniques [85]. Some can be applied during the design, whereas others can be used during operation time of the circuit. A reasonable extract of these techniques is shown in Table 3.1.

Design leakage reduction	*Static leakage reduction*	*Active leakage reduction*
Dual-V_{th}	Stacking	DVS
Multi-V_{dd}	Sleep mode	
	VTCMOS	DVTS

Table 3.1: Leakage reduction techniques

Energy efficient circuits should feature multiple supply voltages and at least a dual threshold approach. As shown in Table 3.1, these characteristics need to be added during the development phase. Furthermore, additional techniques working during operation of the circuit can help to continuously reduce the overall power consumption. Dynamic (threshold) voltage scaling (DVS & DVTS), as well as variable threshold CMOS (VTCMOS) circuitry are powerful methods to overcome the side-effects like subthreshold leakage due to progressive scaling to smaller technology nodes. Again, not all of these methods can be applied without additional arbitrary signals, so a reasonable selection was done for each circuit type.

Chapter 4

Cryptography and its application

4.1 Basic principles

Classic aspects of circuit design like maximum speed, transistor count, area consumption and low-power aspects come usually first when a new development is kicked-off. Dealing with the typical trade-off between these factors is a daily challenge of many engineering teams and must be decided upon the requirements of the customer or the market respectively. Designing an integrated circuit trades off the freedom of choice between engineering out of pre-designed standard cells, semi-custom and full-custom design and layout. Despite the fact that these methods are just an extract of many other possibilities, it can still be considered as the regular state of the art in chip design for the vast majority of target applications.

4.2 Side channel attacks

Whilst encryption became of crucial importance for many applications where sensitive data must be processed / transferred / exchanged, the search for different kinds of attacks on integrated circuits was elaborated in the past years. Generally spoken, three major types of attacks can be identified:

Invasive: An attacker has direct, physical access to the chip and can even re-engineer the internal chip structure without depending on the pins. This is certainly the strongest attack type but it also requires expensive equipment, e.g., for circuit modification by using ion-beams or lasers [118]. The higher the requirements for a successful attack are, the less dangerous it is.

Semi-invasive: Access to the chip is still mandatory as the enclosure must be removed, similar to the invasive attack described before. But the difference here is that the internal structure is usually not modified. Instead of this, semi-invasive attacks are used to extract data, e.g., reading the contents of memory cells for revealing the secret key of an encryption algorithm.

However, this attack is very close to the invasive one once x-rays or electro-magnetic fields are used to stimulate the circuit [118].

Non-invasive: The strongest benefit of the non-invasive attack is its independence of complex and expensive equipment to carry it out. This attack exploits potential vulnerabilities of a chip just by approaching all thinkable interfaces of it. Non-invasive attacks can be put into two different categories: active and inactive. Whilst the active attack still requires a certain level of external impact on the chip, e.g., modifying the temperature or manipulating the clock frequency, the passive variation of it does not require any direct impact on the attack target. Instead, the passive non-invasive attack tries to sense the electro-magnetic radiation of a chip while its computing data or any kind of voltage spikes on the power lines while the chip is operating. Although measuring the voltage spikes or battery current of a design depends on physical access to the target of evaluation, this attack does not depend on extraordinary equipment and therefore cuts down the related costs. An alternative naming of the non-invasive passive attack is side channel attack.

4.3 Power Analysis

Power analysis attacks were in focus of research activities for many years as they are easy and comparably cheap to perform. A total elimination of leakage current emanation requires high efforts on adding special hardware modifications and therefore it is unlikely that a leakage-free circuit can be taken into consideration when talking about security-related countermeasures. Each amount of dissipated power, regardless of the chip type under observation, might reveal information about the computed data inside the logic. Especially registers implemented in CMOS are weak against power analysis attacks as their switching events heavily depend on the stored data [77]. Thus, choosing a special logic style to mitigate the risk of capturing sensitive data is a reasonable decision. Research work concentrates on two main directions: improvement of side-channel attacks for a better exploitation of potential vulnerabilities and improvement of selected countermeasures against such attacks [76]. The first mentioned direction, improvement of side-channel attacks will not be handled within this thesis, whereas the introduction of a power-analysis hardened register will be introduced in Chapter 6. In order to understand how appropriate countermeasures work, it is important to get familiar with the basic principles of power analysis attacks, which will be described in the next two sections.

4.3.1 Simple Power Analysis

A more simple variant of the DPA is the simple power analysis (SPA). The idea behind this kind of analysis is to directly analyze the power consumption of a device, without having any additional statistical analysis results available. In principle, SPA can be used to identify sub-algorithms as certain operations show a distinct signature in the power dissipation [76], especially when template-based attacks are carried out [78]. If a key is processed bit-by-bit, the encryption algorithm will perform either a squaring and multiplication for a logic 1 and a squaring only for a logic 0. Based on these observations, the secret key can be extracted in case that multiplication can be recognized in the power consumption. For example, this can be observed when elliptic curve cryptography (ECC) is performed [79]. SPA can be also utilized as a pre-processing step for other side channel attacks, so despite the fact that it is less efficient than DPA it should be taken into consideration.

4.3.2 Differential Power Analysis

The key principle of a DPA is measuring the dissipated power of a chip by having knowledge about the applied input data and the computed output data [88]. Figure 4.1 displays a simplified flow of such an attack. A certain, well-known plaintext is applied on the inputs of a chip, e.g., on a smart card, and the dumped out cypher text is evaluated together with the continuously traced power drops.

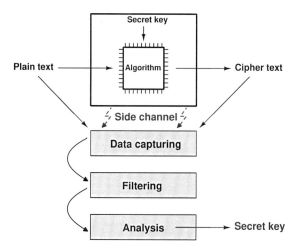

Figure 4.1: Typical approach for a side-channel attack [119]

The technical realization is very simple as it is necessary to place a small resistor between the circuit and *GND* to measure to voltage drops on the resistor. A DPA consists of several steps and the very first one is to get the required knowledge about the applied cipher, which can be found in public literature for popular encryption algorithms. Supportive information about the hardware details, e.g., register size, can be taken into consideration, too. Having done that, the next step is to measure and record the power consumption of the device by using a digital sampling oscilloscope. Special digital signal processors (DSPs) might be required to add an appropriate filtering of the captured power traces. These power traces must be aligned to the hypothetical model, which is the most challenging part. Statistical tests are made to compare the predicted power traces against the measured ones. The more hits are found, the higher the probability will be to reveal secret data. The DPA is finished by applying evaluation algorithms on all captured and measured data to extract secret information out of it. It shall be pointed out here that the success rate of DPAs depends on the complexity of a design and the effort in terms of time and amount of extracted data. With that said, the strength and success of a DPA depends strongly on the leakage model, which must be created in advance and which is used to predict the power dissipation. Different variants of the standard DPA exist, e.g., correlation power analysis (CPA) [80], higher order DPA (HODPA) [81], which can be used to overcome special masking techniques to protect the secret key from being extracted.

4.4 Countermeasures

In parallel to the continuous evolvement of side-channel attacks, countermeasures have been also improved over time. These measures can be applied at different hierarchical level:

Algorithm: Randomizing the execution order at algorithmic level (also known as *shuffling*) is an effective method to make the occurrence of the leakage difficult to predict. Adding random values to intermediate results during computing and subtracting them again after critical operations are completed is known as *masking*. Unless real hardware encryption is available, this type of countermeasure is usually limited to software implementations. On the one hand, this method leads to a higher protection against attacks, but on the other hand it also adds additional efforts to be spend on the software design. Nevertheless, neither *shuffling* nor *masking* will completely remove leakage, so further security-related optimizations should follow.

Architecture: Decoupling the chip from the power consumption / power supply is a reasonable add-on to previous software-based countermeasures. However, as decoupling must still ensure a reliable operation of the chip, further components at hardware level will be required. By going further, transferring the whole design into asynchronous logic or adding more and more different clock domains could be also sensible steps towards higher robustness against DPA. Drawbacks of these techniques are certainly higher costs and more required time for development and verification of circuits with higher complexity.

Logic style: Having covered the algorithmic and architectural level, the last step to be done is also to consider the design of each gate. CMOS logic is the backbone and frequently applied logic style of most chips in use for commercial applications. It is well known and offers decent reliability and speed while pushing down undesired leakage currents to an acceptable level for the vast majority of applications, e.g., smartphones, computers, entertainment electronics. CMOS logic can also be seen as *asymmetric* logic style as the consumed battery current will most likely differ in dependance of the applied input data and the subsequent internal processing of a chip. Furthermore, ASICs and other chips will be either optimized for speed, low-power or area consumption, which also has a direct impact on the layout of the design. In that sense a good layout focuses usually on the support of at least one of these aspects but neglects a well balanced power consumption during each transition. Designing standard cells in a way to counter DPAs, entire technology libraries could be created for the implementation of dedicated circuits of high complexity, which are hardened against any kind of unauthorized attempts. The basic idea behind this attempt is to ensure zero or just minimal variations in power consumption, regardless of the chosen algorithm running on this special logic. Different, appropriate logic styles were under research during the past years and also introduced [82] [83] [84]. Nevertheless, having such libraries available and ready to use also mean to face potential disadvantages which will likely occur due to these improvements that come at cost of certain penalties. These penalties result in increased area footprint and increased power consumption as all applied modifications will lead to complex redesigns of entire libraries. These solution might slow down the operating frequency of a circuit, because a higher delay in signal propagation through complex gates might occur. It should be noted that complex redesign come along with higher costs, which also is a drawback when dealing with special logic styles.

A combination of countermeasures at all these hierarchical levels leads to a good strategy for risk mitigation. As this thesis is focussing on the design of circuits, only potential enhancements / modifications of the logic style for selected circuit parts are in scope of this work. At this point, it is an inevitable step to weight the pros and cons of a special, more complex logic style which is capable of warden DPAs. This topic and its applicability will be handled in detail in Section 6.3.

Chapter 5

Low-power configuration random access memory

5.1 Existing designs

The backbone of each computational activity within an FPGA is the LUT [112]. Typically an FPGA consists of a sea of tiles which contain the necessary logic in terms of LUTs and interconnection circuitry, shown in Figure 5.1. Two different groups of logic can be identified: Configurable Logic Block (CLB) and switch matrix. A CLB is used for ensuring the feature of adaptiveness due to the built-in LUTs, therefore it contains the LUTs and additional components, e.g., flip-flops, multiplexers and basic logic gates. On the other hand, the switching matrix is used for providing all necessary interconnections to other tiles / LUTs in case that more complex functions are requested to be implemented and require a combination of multiple CLBs.

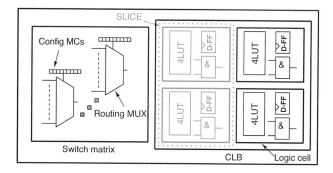

Figure 5.1: Simplified 'Tile' of an FPGA

By putting a higher focus on the optimization of the configuration RAM cells, these efforts to not only improving the power balance of the CLBs, but also to decrease the switch matrices standby

41

leakage currents. For communication with peripheral logic, General Purpose Input Output (GPIOs) blocks are implemented which can be used for bidirectional data. However, switch matrices are not subject matter of this paper and will be discussed in later publications.

Depending on the number of the LUT's inputs, a LUT can contain numerous SRAM cells. For example, in case of a 4-input LUT, 16 SRAM cells are necessary for the realization of all possible input value combinations. An exemplary illustration of a LUT is shown in Figure A.2. Since the memory cells are used for configuration, they are also called configuration RAM (CRAM). Once configured during the start-up phase, the content of these memory cells would not be changed until the next reconfiguration cycle. In consequence, the static leakage current reduction is of higher significance for the overall power consumption. The selection of a low-power SRAM cell design is crucial for an appropriate energy-efficient implementation of integrated circuits. Many memory cell designs have been introduced in the past. In principle, this memory cell consists of two cross-coupled inverters and two access transistors, connecting the inverters to the bitlines, as shown in Figure 5.2 [73] [74].

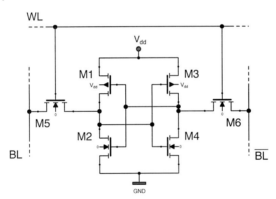

Figure 5.2: 6T SRAM cell

As long as $M5$ and $M6$ are in cut-off mode, the crosscoupled inverters are isolated from the bitlines and store the complementary data value at the output nodes of each inverter. Data retention is ensured as long as a sufficient supply voltage V_{dd} is applied. Before reading the stored data, both bitlines BL and $\overline{BL}$ are precharged to V_{dd} by a special precharge circuit and the access transistors $M5$ and $M6$ are turned on. One of the bitlines will be discharged to GND, whereas the other bitline will remain on V_{dd}. The voltage drop between BL and $\overline{BL}$ will be sensed and evaluated by a sense amplifier. For writing data into the cell, one of the bitlines is kept at V_{dd}, whereas the other bitline is kept at GND. By turning the access transistors on, the desired value is written. For this purpose, a suitable bitline driver circuit is needed to ensure the proper execution of the writing cycle. Careful transistor sizing is required for avoiding the cell to flip during, e.g., a read cycle. This cell design is well-elaborated and used for years in integrated circuits. Its stability and reliability is well-known [98], [107], [120] and therefore used in various applications. However, the power consumption of the 6T SRAM cell can be further optimized by some modifications resulting in the SRAM cells described in the following paragraphs:

4T SRAM cell

A typical implementation of a four transistor SRAM cell is shown in Figure 5.3. In comparison to the 6T cell, a smaller area of approximately 30% can be achieved [116]. Due to the replacement of all pMOS transistors by polysilicon resistors, only nMOS transistors are used for the pure functionality of this cell. Despite of the space-savings, which could lead to a higher yield after the manufacturing process, the realization of high-resistivity polysilicon resistor adds additional technological steps to the manufacturing process, resulting in higher costs.

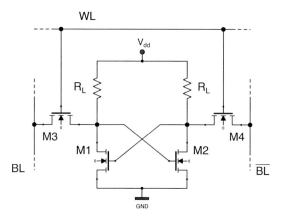

Figure 5.3: 4T SRAM cell

Furthermore, replacing pMOS transistors by a resistive load means to disclaim the advantages of CMOS technology in terms of leakage current suppression. As pMOS transistors act as switches by cutting off the circuit from V_{dd} when used in pull-up networks (PUNs), this benefit gets lost if resistors are used instead. This can be seen in Figure 5.4.

43

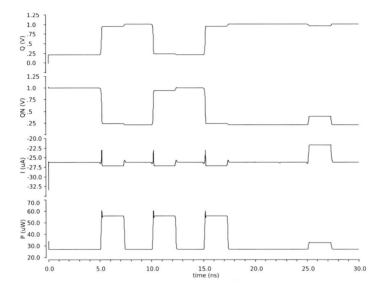

Figure 5.4: Power dissipation P and power supply current I of 4T SRAM cell

Operating this memory cell design with full-swing output requires two resistors in the PUN of $30k\Omega$ each. Everything less than $30k\Omega$ leads to a degradation of the complementary output nodes Q and $\overline{Q}$ by not achieving full rail-to-rail output. Based on the curves in Figure 5.4 the correct function of this cell is verified, however, this figure also shows a continuous current I flowing through the circuit even without switching activities. That also correlates with a higher power consumption P compared to the output curves of the other designs, which will be introduced in the following sections.

$\varnothing$ P (μW)	max$\{P\}$ (μW)	min$\{P\}$ (μW)	$\varnothing$ I (μA)	max$\{I\}$ (μA)	min$\{I\}$ (μA)
33.13	59.81	26.21	-26.13	-21.61	-33.33

Table 5.1: 4T SRAM cell simulation results

All measured results in Table 5.1 support the previous statement about a high power consumption in comparison to the other investigated designs. In order to complete the investigation of the overall performance of a memory cell, the static noise margin (SNM) for read operations and write noise margin (WNM) for write operations should be also taken into consideration [94]. SNM is defined as a dedicated voltage level of noise on the bitlines, which may come up but will not be able to flip the stored bits during a read process. Hence, the bigger SNM is the more stability is provided by the respective memory cell design. A SNM of at least $200mV$ should be achieved to ensure sufficient stability of the cell in read mode [110]. Figure 5.5 shows the SNM of the 4T SRAM cell.

44

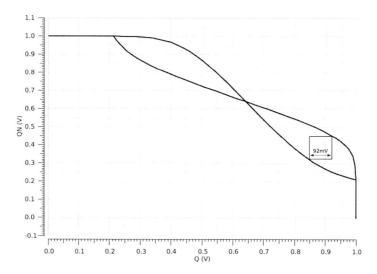

Figure 5.5: SNM evaluation of the standard 4T SRAM cell

The value of the SNM is the side of the largest square fitted in one of the lobes of the *butterfly curve*. Here, the read SNM is $92mV$ only, which is a weak read stability of the cell. For a better result, the cell ratio (CR) comes into play at this point, which is defined by Equation 5.1.

$$CR = \frac{W_1/L_1}{W_3/L_3} = \frac{W_2/L_2}{W_4/L_4} \tag{5.1}$$

So basically, the (symmetric) CR is defined as the relation of the transistor sizes in the pull-down network (PDN) compared to the access transistor size. The higher the cell ratio is, the better read stability will be. In consequence, PDN transistors should be stronger than access transistors, hence, they should have a bigger width. On the one hand, a larger cell ratio makes the memory cell robust and provides higher SNM and read current, which is a precondition for a higher operating speed. On the other hand, increasing the width of a transistor means also to increase parasitic capacitances and area consumption. Especially parasitic capacitances are an undesirable effect as the primary goal is to mitigate power dissipation to a minimum level. In consequence, the decision was taken to leave the transistor size at the minimum size, at least as long as correct functionality is not affected. Another aspect for consideration is the WNM, which is shown in Figure 5.6 for the 4T SRAM cell.

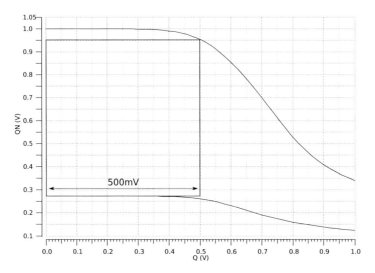

Figure 5.6: WNM evaluation of the standard 4T SRAM cell

The WNM is measured by sweeping the voltage at the storage node Q with BL and WL stuck at V_{dd} and $\overline{BL}$ is connected to GND while monitoring the voltage at node $\overline{Q}$. Figure 5.7 illustrates the respective measurement setup.

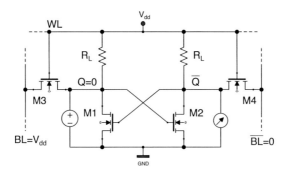

Figure 5.7: WNM measurement setup for the 4T SRAM cell

Similar to the previous SNM measurement, the WNM can be extracted by fitting the largest side of a square, as shown in Figure 5.6. The 4T SRAM cell shows a comparably decent result WNM result, but strongly depends on the choice of R_L. The reason for this outcome is that either $M3$ and R_L or $M4$ and R_L form a resistive voltage divider during one of the falling bitlines. Transistor sizing plays also a role at this point, leading to the definition of a pull-up ratio (PR) which is the counterpart to the CR of SNM. For example, the (symmetric) PR of the design shown in Figure 5.2 is defined by Equation 5.2.

$$PR = \frac{W_1/L_1}{W_5/L_5} = \frac{W_3/L_3}{W_6/L_6} \tag{5.2}$$

All of these investigations have also been executed for the 6T SRAM implementation and are shown in the following illustrations. At first, the static noise margin was measured and can be found in Figure 5.8.

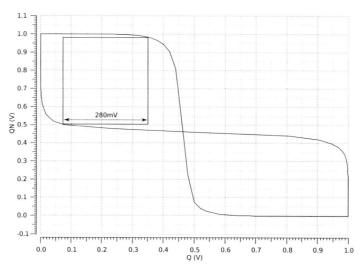

Figure 5.8: SNM evaluation of the 6T SRAM cell

A direct comparison of both SNM values underlines that a 6T SRAM cell inhibits a better read stability by having a SNM of $280mV$, which is approximately 3 times higher than before. This goes back the stabilizing looped inverters between the access transistors. Moving forward means to consider the write noise margin as well, therefore the WNM result can be seen in Figure 5.9. A comparison of both evaluations shows that the write noise margin of the 6T SRAM cell is slightly worse than of the reference design, resulting in a difference of $75mV$.

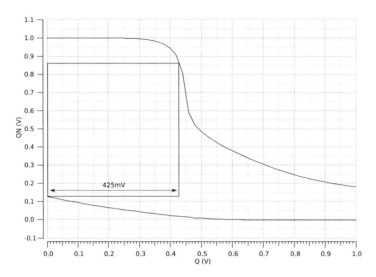

Figure 5.9: WNM evaluation of the standard 6T SRAM cell

In summary, it can be stated that the standard 4T (polysilicon) SRAM cell is a predecessor of all CMOS-based SRAM cells. Due to the constraints in terms of additional process steps and costs in manufacturing, this design does not play a major role in modern memory arrays any more. Lower stability, lower tolerance against soft-errors and a more technically demanding manufacturing process exclude this cell type from further considerations [117].

4T SRAM Noda loadless cell

The biggest downside of the standard 4T SRAM cell is the resistive load in the PUN, which adds some undesired disadvantages when thinking about the overall performance of that cell. So by recognizing the root cause of these disadvatages, the corollary is to replace these resistors by something else or to eliminate them completely. The circuit shown in Figure 5.10 introduces a thinkable solution [110]. The former nMOS transistors have been replaced by their pMOS counterparts. All resistive load elements were removed from the original design. During a standby phase, the access transistors act as a load element to the driving nMOS transistors in the PDN network. Furthermore, by using resistors in the PUN, a voltage drop V_{R_L} over these resistive elements will always occur. By removing all R_L elements, this problem is bypassed and the circuit's characteristics are improved. First of all, the cell was placed in a testbench and simulated to verify its correct operability (Figure 5.11).

48

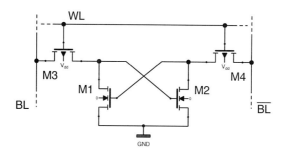

Figure 5.10: 4T SRAM Noda cell

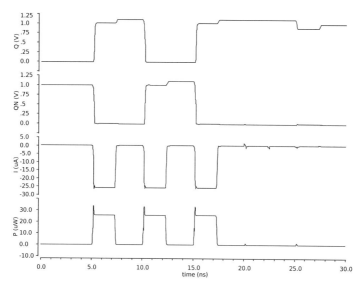

Figure 5.11: Power dissipation P and supply voltage current I of 4T SRAM Noda loadless cell

Discussing the 4T SRAM Noda cell also means to take a look of SNM and WNM for having a comparison of the previously investigated reference design. Figure 5.12 depicts the *butterfly curve* of the 4T SRAM Noda cell. The best fitted square in one of the butterfly lobes reveals a SNM of $155mV$, which is an obvious improvement when compared to the $92mV$ SNM of the original design. Nonetheless, it should be stated at this point that the SNM strongly depends on the transistor parameters, as the SNM can get a decent improvement like shown in Figure 5.13. All shown colors of the displayed curves represent different widths of the pMOS transistors ($120nm$ up to $720nm$) inside the cell.

49

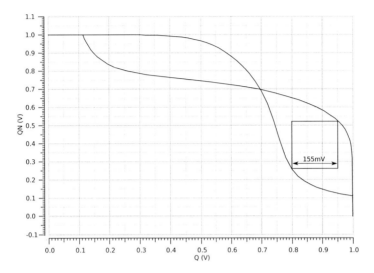

Figure 5.12: SNM evaluation of the 4T SRAM Noda loadless cell

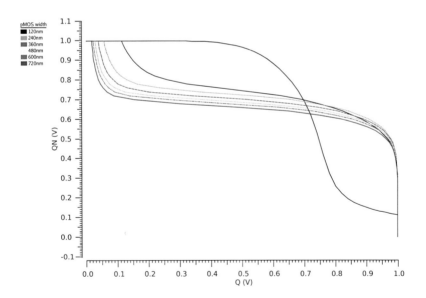

Figure 5.13: SNM dependence of CR

Despite the potential improvements in terms of achievable static noise margin, a width of $720nm$ is six times wider than the original width of $120nm$. Here it comes again to the trade-off between low power aspects and SNM. By still having power savings as main goal ahead, minimum transistor sizing was used as basis for further comparisons. Similar to the previous investigations for the standard 4T SRAM cell, the WNM for the 4T SRAM Noda cell is highlighted in Figure 5.14. Although the WNM is slightly worse than the respective result of the previous design, it is still considerable by showing write noise margin of approximately $430mV$. The overall result of the Noda cell is better and represents an acceptable improvement over the reference design.

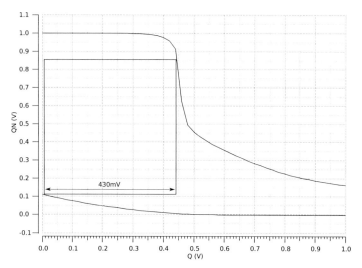

Figure 5.14: WNM evaluation of the 4T SRAM Noda loadless cell

51

5T SRAM cell

A variation of the previous designs is the five transistor SRAM cell, shown in Figure 5.15. The basic idea behind this modification is to decrease area consumption by redesigning the typical layout of a memory cell.

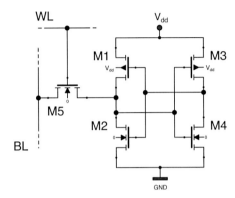

Figure 5.15: 5T SRAM cell

The advantage of this cell design compared to the 6T reference cell is the availability of just one access transistor $M5$ and therefore only one bitline BL [101]. The connecting bitlines in each slice of an FPGA add undesired parasitic capacitances, which underly the process of charging and discharging during each read- and write-cycle and lead subsequently to higher power consumption. A cell design working with just one access transistor adds space-savings. For a proper and stable functionality of this cell, asymmetric transistor sizing is required, which may complicate the manufacturing process and lead to modifications of auxiliary circuitry like sense amplifiers, precharge circuits, etc.. Hence, the underlying concept of this design is to decrease the trip-point of the right inverter ($M3$ and $M4$) and to increase the trip-point of the left inverter ($M1$ and $M2$). In addition to that, the access transistor requires careful sizing to be used for both, read and write operations. A benefit of the asymmetric cell sizing is the lower precharge voltage on the bitline, which is required to run a stable, non-destructive read operation. Simulations have shown that a precharge voltage V_{pc} of approximaltey $500mV$ is sufficient to operate read-cycles under stable working conditions. For evaluating how asymmetric sizing may impact the static and write noise margin, the respective simulations were executed and their results are shown in Figure 5.16 and Figure 5.17. A comparison between the 4T SRAM Noda cell and the 5T SRAM cell shows, that the results of the 5T design are slightly better balanced than of the previously described implementation. Here, both benchmarks are very close to each other (SNM $250mV$ and WNM $300mV$), whereas the Noda cell's difference between SNM and WNM is $275mV$. One reason for this circumstance is the special adaption of all transistor parameters within the circuit and therefore a well balancing of read capabilities and writing strength. Despite these benefits, it should be mentioned

52

here that increased transistor sizes are equal to higher power supply current. This is an undesirable effect, which needs to be mitigated or, in the best case, avoided.

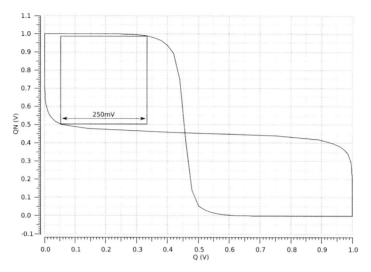

Figure 5.16: SNM evaluation of the standard 5T SRAM cell

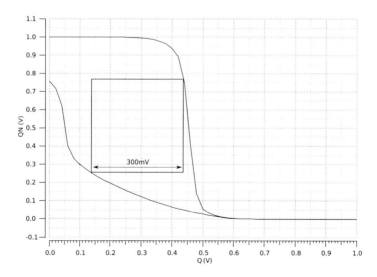

Figure 5.17: WNM evaluation of the 5T SRAM cell

53

7T SRAM cell

The seven transistor SRAM cell is shown in Figure 5.18, which enhances the 6T reference cell design by an additional feedback transistor $M7$ and 2 signal lines R and W. The idea behind this design is a write mechanism, which depends only on one of the two bitlines in order to execute a write operation. This can be also expressed in Equation 5.3 [1].

$$P = \alpha C_{BL} V_{dd}^2 F_{write} \tag{5.3}$$

Whilst the activity factor α equals 1 in conventional memory cells, the 7T SRAM cell reduces this factor to less than 0.5 by exploiting the fact, that most of the bits in memories and caches are zeros [111]. C_{BL} brings the parasitic bitline capacity into this equation and F_{write} the frequency of write access events. The main asset of this implementation is the reduction of the switching activity and therefore a reduction of charging and discharging cycles of parasitic capacitances. The drawback is the required additional control logic and the loopback transistor, which lead to higher complexity and required space. The basic idea of this design is a special characteristic of the write process.

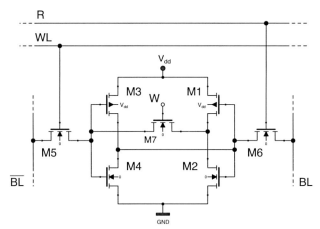

Figure 5.18: 7T SRAM cell

At first, the feedback transistor $M7$ is turned off by applying *HIGH* at its gate connector, which cuts off the feedback loop. The consequence now is that the core of the 7T SRAM cells is transformed into two cascaded inverters, INV1 and INV2 shown in Figure 5.18. Second, access transistor $M5$ is activated to transfer the data from $\overline{BL}$ to $Q2$, driving the input of first inverter INV1 in the line of the cell core. As consequence Q is created on the node between both inverters, depicting the true cell data. After that Q drives the next inverter INV2 in the data chain to generate Q, which equals $Q2$. Storing the created data inside the cell is ensured by setting the wordline WL to LOW and setting W to $HIGH$. $N5$ is turned on and serves to reconnect the feedback loop between INV1 and INV2. Figure 5.19 shows an illustration of this principle.

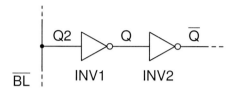

Figure 5.19: Write operation

Both bitlines are precharged to V_{dd}. Following the described write process, $\overline{BL}$ is kept at V_{dd} to store a 0 into the cell, coming along with a negligible power consumption due to decreasing the activity factor α below 1. The simulation results of this cell are illustrated in Figure 5.20.

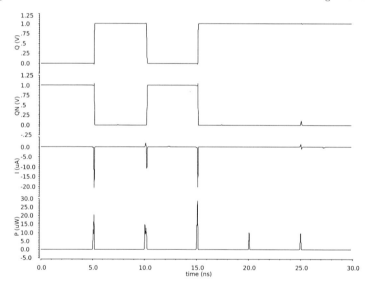

Figure 5.20: Power dissipation P and power supply current I of 7T SRAM cell

Before moving on to the next chapter, it is still of interest to put a spot on SNM and WNM results of this cell. The result of SNM, depicted in Figure 5.21, does not show any surprising details on its track. It is mostly comparable to the 5T SRAM cell and is an acceptable result. Coming to the next result, write noise margin, a noticeable improvement over the WNM of the 5T SRAM design can be seen in Figure 5.22. Here, the stability and better balanced write capabilities go back on two bitlines instead of having one. For a better overview, all determined results until this point are summarized in Table 5.2.

SRAM type	4T	4T Noda	5T	6T	7T
SNM (*mV*)	92	155	250	280	270
WNM (*mV*)	500	430	300	425	415

Table 5.2: SNM and WNM comparison

Having these results in mind, it can be stated that the 6T SRAM cell holds the best balanced results in terms of noise margin for both, write and read operations. In addition to that, each analyzed design had to undergo a transient analysis to take a closer look into its dynamic behavior and to extract the power dissipation during runtime.

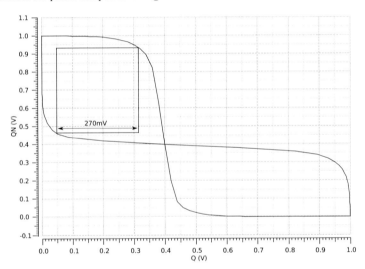

Figure 5.21: SNM evaluation of the 7T SRAM cell

56

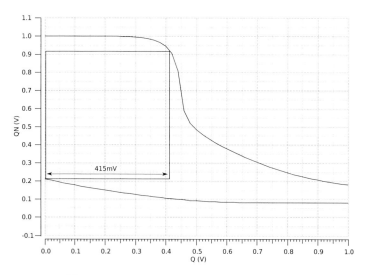

Figure 5.22: WNM evaluation of the 7T SRAM cell

These results will be highlighted in the next chapter as they will serve as a potential benchmark for the overall performance of the new low-power SRAM cell.

Test results

All SRAM cells have been designed and simulated by usage of the Cadence toolchain and a $90nm$ technology provided by TSMC at an ambient temperature of 27°C. The main challenge to achieve comparable results was to develop suitable bitline drivers, precharge circuitry and a sense amplifier. Careful design of the bitline drivers is crucial for avoiding the cell to flip during a read cycle. All simulations were performed by applying a clock frequency of $200MHz$ and a load consisting of a sense amplifier, connected to each bitline. According to the product specification of the Xilinx Spartan 3A FPGA, the internal DCMs cover a frequency range from $5MHz$ to $320MHz$ [22]. Due to this information, a clock frequency of $200MHz$ was selected as it depicts an acceptable trade-off within the given range. The configuration memory cells used in a LUT are not supposed to be written and read at high frequencies, like, e.g., memory arrays in a microprocessor's cache, which might operate up to $4.2GHz$. Therefore, we choose a lower frequency, nevertheless all cells have also been successfully tested with a higher clock frequency of $500MHz$, for making sure that these cells will also work well out of the intended specification. As expected, the higher operating frequency led to higher power consumption. All cell designs have been applied to the test circuit in Figure A.1. This shows a 6T SRAM cell as DUT (design under test), the precharge circuit consisting of transistors $M7$, $M8$ and an equalizing transistor $M9$, two bitline drivers ($M15$, $M16$ and $M17$, $M18$) and a sense amplifier. For the first step, the determination of the best SRAM cell design in terms of power consumption without any further improvements, is done. The simulation results of the 6T cell design are shown in Figure 5.23.

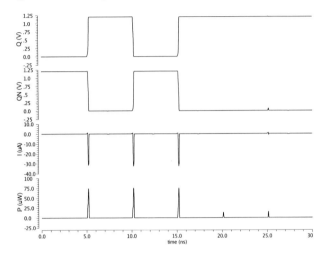

Figure 5.23: Power dissipation P and power supply current I of 6T SRAM cell

The average power consumption, the maximum and minimum power consumption during simulation time were traced and summarized in Table 5.3.

$SRAM\ cell$	$\varnothing\ P\ (nW)$	$\max\{P\}\ (\mu W)$	$\min\{P\}\ (pW)$
4T Noda	334.5	35.07	161.7
5T	587.2	61.26	217.34
6T	927	75.39	250.8
7T	491	49.19	221.7

Table 5.3: Simulation results without modification

Compared to the other designs, Table 5.3 demonstrates the drawbacks of the reference 6T SRAM cell. Substantial power savings can be achieved by the choice of alternative cell design. For example, the average power consumption of the 6T SRAM reference cell design is $927nW$ and about 3 times higher than the average power consumption of the 4T SRAM Noda loadless cell, which is only $334.5nW$. That results in power savings of approximately 65%.

Among these proposed cells, further designs have been presented with focus on low-power applications. The proposed 7T design was slightly modified by employing an improved self controllable voltage level (ISVL) circuit [56]. By adding this auxiliary circuit, full V_{dd} shall be supplied in active state, whilst switching to a lower supply voltage in idle state. Simulation results have proven the feasibility in terms of decreasing static power dissipation. The drawback here is that these extensions are capable of lowering static power consumption only, despite the addition of seven transistors. With the dawn of IoT devices, an 8T SRAM cell was introduced and designed with special focus on operation in sub-threshold or near threshold voltage of CMOS logic [57]. This cell can be powered by very low supply voltages of approximately $400mV$ while still providing a sufficient SNM for reliable read access. On the other hand, the maximum circuit speed is limited and high V_{th} transistors are still required. Therefore this design might be suitable for ultra low-power applications where operating frequency is a minor concern. By trying to keep the number of transistors as low as possible but to achieve better results for extended battery lifetime, a SRAM cell with carbon nanotube field effect transistors (CNFETs) was published [58]. The idea was to keep circuitry for read and write operations separated for optimized access times while maintaining low-power characteristics. CNFETs are responsible for avoiding high power consumption for both, static and dynamic considerations. Nevertheless, the success of this design strongly depends on reliable CNFET manufacturing and operation as well as careful adaption of the sense amplifiers, which contributes most to the improved operation speed.

Another research work focussed on elaborating cell extensions for leakage reduction by cutting off supply voltage and *GND* from the cell core as good as possible [59]. The strategy of this focus work goes back to dynamic voltage scaling, which is applied to the supply voltage node of the memory cell. In addition to that, the output node of a supply voltage controlled inverter is connected to the virtual *GND* note of the memory cell. This modification shall support dynamic adaption of the V_{dd} and dynamic regulation of the cell's virtual *GND* node voltage level. Here, the most significant drawback is the requirement for a dedicated V_{dd} controller circuitry, which consists of eight transistors. By considering the original six transistors of the core cell and the additional two transistors of the inverter in the *GND* path, this implementation requires 16 transistors in total. Based in that, it can be stated that the complexity of this design must be carefully compared against achievable power savings. The introduction of a new component into SRAM cell design was also published in the recent years [60]. Memristors are non-linear resistors and are able to act as memory with improved power and speed characteristics. If a cell design is built upon a combination of memristors for data storage and multi-threshold CMOS for gating functions, it can be a remarkably good solution for applications with constraints in terms of available energy resources. Memristor technology was invented in 1971 and successfully manufactured in 2008, however, memristor enhanced memory cells show-up disadvantages in circuit speed when compared to competing designs and require additional manufacturing process steps.

A further enhanced design, the new 9T SRAM cell, was presented as result of other research activities [61]. This cell uses nine tunnel field effect transistors (TFETs) and had efficiently suppressed leakage currents and provided an improved read stability. For this purpose, an additional, dedicated read bitline was added. The downside of this design is the increased number of transistors, without adding common power reduction techniques. However, the transistor type necessary for this technique has direct impact on the manufacturing technology by adding additional process steps. In the end, similar to the analyzed 7T SRAM cell, the additional bitline leads to the undesired consequence of increased routing effort and parasitic load. Similar to the introduced 5T SRAM cell in Section 5.1, an alternative, single-ended 10T memory cell was the outcome of research [62]. In contrast to the 5T SRAM cell, this design inhibits a different architectural approach for further power savings and increased read / write stability even at low supply voltages. Thus, the original design was doubled in number of transistors and transmission gates were introduced. Despite the fact of representing a single-ended memory cell, this design requires additional control signals for proper operation, which detracts the advantage of single-ended architecture and exhibits a complexity without special attention to low-power optimization. An interesting approach is to optimize read cycles of a SRAM cell by skipping the *precharge* process of the bitlines [63]. This is realized by extending the classic 6T SRAM circuit by an additional inverter and transmission gate. The inverter is used as readout inverter and able to fully charge and discharge the read bitline, hence a dedicated *precharge* scheme is not required. The drawback of this design is the necessity of two additional bitlines, which must be taken into consideration.

An alternative design is the 12T SRAM cell, which features dual-port access for increased access times while still keeping low-power aspects in mind [64]. Dual-port SRAM cells provide a performance increase based on faster access mechanisms, qualifying these cell types for the support of high performance applications such as image processing in biomedical devices. This special characteristic allows operation at comparably low supply voltages, however, read- und write stability gets decreased compared to other designs. A baseline number of 12 transistors is disadvantageous for further enhancements with power reduction measures and a total number six access lines leads to introduction of potential parasitic capacitances. Closely allocated to that is the requirement for special control logic, decoders, etc. Further research was spent on the elaboration of a good balance between low-power consumption, speed and area footprint [65]. A series of different SRAM designs (8T, 11T, 13T) was designed based on dynamic logic. This technique allows higher circuit speed, especially when combined with differential outputs. However, as SRAM cells shall be mostly used as configuration memory for LUTs, highest achievable switching frequencies are not of first priority for the intended applications.

5.2   New low-power approach: LP 4T SRAM Cell

The previous simulations of selected legacy cells have shown that the choice of a suitable SRAM cell design leads to a significant impact on power consumption of a LUT. Major goal of this research was to develop a memory cell, which features best battery charge sustainment even if other attributes like SNM, WNM or maximum operating frequency may suffer. The first step to do so was to choose a cell design which comes along with minimized power dissipation from scratch but still provides enough room for further improvements. In parallel to these activities, the same improvements were applied on each legacy cell design for checking the improvements of the new cell over the existing ones. Since Xilinx' Spartan 3(A) is manufactured in a $90nm$ process and has a recommended internal supply voltage range from $1V$ to $1.2V$, we choose a $90nm$ TSMC technology library at a comparable operating voltage of $1V$. As the consumption of energy is tightly coupled to the supply voltage V_{dd}, it is a reasonable approach to keep this factor as low as possible.

Coming back to the proposed cell designs in Section 5.1, the 4T SRAM came into focus, since its compact design is of interest for further considerations and performance comparison to other designs. But on the other hand, a major drawback of the 4T SRAM cell is the high-resistive polysilicon resistor, which should be replaced or completely removed in an alternative, improved cell. An option for such an alternative to this implementation is the introduced 4T SRAM Noda cell, which eliminates this drawback by using the access transistors as load. So the question came up whether there is a chance to redesign this cell in a way to save energy from the very first concept idea. Bitlines are usually precharged to V_{dd}, which drains energy from the battery and can not be avoided in standard designs. Therefore the new low-power 4T SRAM cell features a *precharge* process to *GND* instead of V_{dd}. For a correct operation of the intended cell, minor adaptions have to be applied to the internal structure by swapping the nMOS with the pMOS transistors as compared to the Noda cell. This means that the previous pull-down network (PDN) consisting of two nMOS transistors is replaced by a pull-up network of two pMOS $M1$ and $M2$ transistors. The underlying principle behind this idea is shown in the circuit of Figure 5.24.

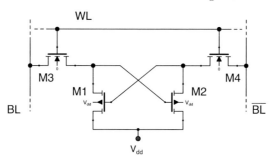

Figure 5.24: 4T loadless SRAM cell

In combination with both - nMOS access transistors $M3$ and $M4$ - a stable and power saving functionality is achieved. Instead of precharging both bitlines to V_{dd} as a pre-step for the following reading-phase, the bitlines are "precharged" to GND, due to the fact that pMOS transistors are used as drivers in this cell. This saves power and ensures compatibility with CMOS logic processes. Nevertheless, minor adaptions to the auxiliary circuitry around the cell have to be done, e.g., modifying the bitline drivers.

5.2.1   Dual threshold CMOS

Further optimizations can be achieved by the introduction of high threshold voltage (V_{th}) transistors. High V_{th} transistors require a higher V_{gs} voltage at the gate in order to turn the transistor on, which can lead to an increase of the propagation delay within a signal path. Therefore, high V_{th} should be only used in applications which are not timing-critical. However, the SRAM cells in a LUT are used as configuration RAM (CRAM) and are pertinent for use with high threshold voltage transistors. All cell designs have been modified and the simulations were performed again. These modifications are limited to the core cell only; the precharge circuitry, the sense amplifier and the bitline drivers have not been modified. The results are summed up in Table 5.4.

$SRAM\ cell$	$\varnothing\ P\ (nW)$	$\max\{P\}\ (\mu W)$	$\min\{P\}\ (pW)$
LP 4T hvt	324	21.27	49.59
5T hvt	541.78	25.1	189.4
6T hvt	695.1	44.67	166.1
7T hvt	427	26.53	167

Table 5.4: Simulation results with high threshold voltage transistors (hvt)

In comparison to the reference design of the 6T SRAM cell, the introduction of the high V_{th} transistors adds power savings of about 25%. The performance of the high V_{th} 4T loadless SRAM cell is slightly improved and leads to energy savings of approximately $10nW$. In general, we can say that this modification improves both, the maximum and minimum energy consumption of all introduced cells. For illustration purposes, these improvements are shown in Figure 5.25. Despite the energy savings which come along with the usage of high V_{th} transistors, the LP 4T SRAM cell shows its superior reduction of consumed power just by its initial design. This will be even more improved by the following steps. Since the 4T loadless SRAM cells offers an excellent out of the box power balance, the strongest impact of the first optimizing steps can be primarily noticed in all already existing designs, especially the 6T SRAM cell. Independent from the respective implementation, each exploited measurement is positively influenced by the modification done in a step before.

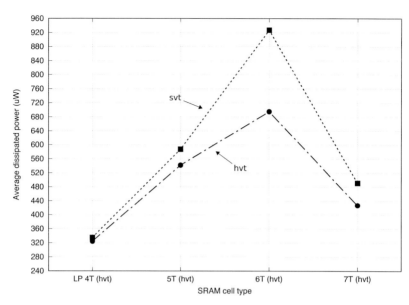

Figure 5.25: Comparison of average PWR dissipation between standard (svt) and high (hvt) V_{th} designs

5.2.2 Transistor stacking

Transistor stacking, shown in Figure 5.26, is a strong technique to reduce subthreshold leakage current by raising the voltage at the source terminal of each transistor.

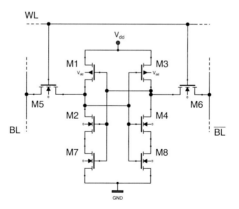

Figure 5.26: 6T SRAM cell with stacking

By constantly increasing the source voltage V_s and keeping the gate voltage V_g at the same level, V_{gs} becomes negative at a certain point of time, which leads the transistor into super cut-off mode and turns it deeply off. Subthreshold currents are exponentially reduced.

At the same time, the body to source potential V_{sb} also becomes negative, since the body terminal of a nMOS transistor is usually kept at *GND*. In consequence, the body effect is intensified, thus V_{th} is tuned by that effect to a higher level. This can be further exploited by continuing stacking transistors in series; it must be noted that the effect of subthreshold current reduction becomes diminished with a rising number of transistors. This technique implies a trade-off between power savings and size ratio of the chip. Despite the gradual technology shrink up to $16nm$ FinFET, on-chip space is not an unlimited resource and should be used carefully. Therefore, it was decided to add only two stacking transistors in order to have a reasonable compromise between leakage current reduction and size-ratio of the cells. The simulation results are shown in Table 5.5 and Table 5.6.

SRAM cell	$\varnothing$ P (nW)	max$\{P\}$ (μW)	min$\{P\}$ (pW)
LP 4T	346.8	35.31	137.6
5T	561.2	26.8	212.9
6T	826.6	72.05	274
7T	540.4	31.64	168.3

Table 5.5: Simulation results with standard transistors and stacking

If the used manufacturing process does not support dual-threshold CMOS technology, Table 5.5 shows that a noteworthy reduction of leakage currents within the 4T SRAM cell is achieved by approximately 90%. Even the standard 6T SRAM cell features important amendments in terms of power savings (approximately 12%) and leakage currents.

SRAM cell	$\varnothing$ P (nW)	max$\{P\}$ (μW)	min$\{P\}$ (pW)
LP 4T hvt	336.6	32.79	70.42
5T hvt	327.4	24.3	128.2
6T hvt	672.4	61.28	167.4
7T hvt	461.8	30.84	523.9

Table 5.6: Simulation results with *hvt* transistors and stacking

In Table 5.6, the result for the average power dissipation of the LP 4T SRAM cell is higher than prior the application of transistor stacking, shown in Table 5.4. The reason for this is that transistor stacking is a technique used for static leakage reduction as described in Table 3.1. Based on that, the most significant contribution of transistor stacking can be seen best when analyzing static leakage current, shown in Figure 5.27.

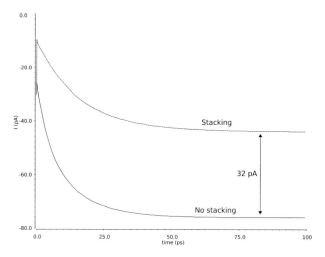

Figure 5.27: Leakage current reduction by applying stacking in the 6T SRAM cell

Figure 5.27 displays the positive effect of stacked transistors in the PDN of a 6T SRAM cell. The upper output curve depicts the leakage current flow through the memory cell after switching off the access transistors and stopping any write or read activities. Hence, the circuit remains in a standby mode without being switched off from the power supply. At this point, the additional transistors in the PDN help to suppress undesired currents. A short comparison of the stacking effect in the LP 4T and 6T SRAM cells highlights the difference, shown in Table 5.7 and in Table 5.8.

Technique	$\varnothing\ P_{LP4T}\ (pW)$	$\varnothing\ I_{LP4T}\ (pA)$	$\varnothing\ P_{6T}\ (pW)$	$\varnothing\ I_{6T}\ (pA)$
No stacking (svt)	76.5	$\lvert 65.29 \rvert$	118.4	$\lvert 75.49 \rvert$
Stacking (svt)	61.06	$\lvert 59.28 \rvert$	85.43	$\lvert 43.41 \rvert$
Difference (%)	20.2	9.2	27.85	42.5

Table 5.7: Stacking effect on static leakage current suppression of the LP4T and 6T SRAM cell based on *svt* transistors

Technique	$\varnothing\ P_{LP4T}\ (pW)$	$\varnothing\ I_{LP4T}\ (pA)$	$\varnothing\ P_{6T}\ (pW)$	$\varnothing\ I_{6T}\ (pA)$
No stacking (hvt)	13.37	$\lvert 11.8 \rvert$	14.17	$\lvert 18.14 \rvert$
Stacking (hvt)	10.32	$\lvert 10.54 \rvert$	11.07	$\lvert 14.95 \rvert$
Difference (%)	22.88	10.68	28.93	17.59

Table 5.8: Stacking effect on static leakage current suppression of the LP4T and 6T SRAM cell based *hvt* transistors

Table 5.7 and Table 5.8 show the benefit of transistor stacking as it mitigates undesired leakage current flow through the memory cell in idle state. This also has a positive impact on power dissipation in idle state, even when high V_{th} are provided by the process technology and already implemented in a design. A negative impact of transistor stacking might occur in dynamic power dissipation as each additional transistor adds parasitic capacitances which will be charged and discharged during runtime, hence the slightly higher dissipated power in Table 5.6. Nevertheless, designing a low-power optimized memory cell requires to consider both, static and dynamic power dissipation. In consequence of that, the combination of both techniques, dual-threshold CMOS and transistor stacking, puts additional improvements to the overall power savings parameters. Since most of the currently available technologies feature dual-threshold CMOS, the feasibility of this combination is high.

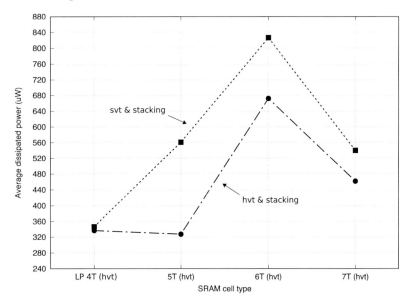

Figure 5.28: Comparison of average power dissipation between standard and high V_{th} designs with stacking

Figure 5.28 display different characteristics of the consumed power by referring to the values of Table 5.5 and Table 5.6. The 5T SRAM offers a slight advantage compared to the LP 4T SRAM design due to higher active and standby intrinsic power consumption of less transistors when applying stacking to a logic design. Nevertheless, the LP 4T SRAM cell still performs better than the remaining memory circuits. Another interesting aspect to be considered is the signal to noise ratio, which gives a benchmark about the margin between the transferred signal or stored data inside a memory cell and the influence of background noise on the signal lines, which can not be neglected. This factor is even of higher significance when volatile memory cells are equipped with high V_{th} transistors, replacing their standard V_{th} counterparts.

5.2.3 Dynamic voltage scaling

The higher the supply voltage is, the faster the operation of the integrated circuit will be, since high V_{dd} allows fast charging and discharging of parasitic capacitances. In case of low demand on performance such as for CRAMs, the supply voltage can be lowered while still ensuring data retention within the cell. Dynamic voltage scaling (DVS) depends usually at least on an operating system and a regulation loop to recognize the circuit speed and to cover a wide range of operating voltages. The proposed approach of this research work simplifies this principle by introducing two additional transistors, shown in Figure 5.29.

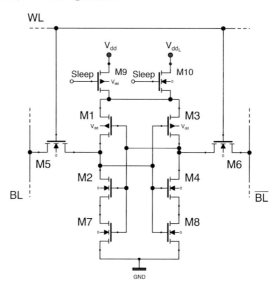

Figure 5.29: 6T SRAM cell with *hvt* transistors, stacking and DVS

Both transistors $M9$ and $M10$ are used to connect the SRAM cell to two different supply voltages, V_{dd} and V_{dd_L}, whereas V_{dd} equals the primary $1\,V$. On the one hand, the prerequisite of this method is a dual-V_{dd} setup, representing a simple alternative to the mentioned operating system driven regulation loop, and on the other hand, a modified power gating approach is implemented. Since the 4T SRAM cell has no connection to GND in its core, power gating is achieved by the possibility to fully cut-off the supply voltage, if needed. However, power gating should be introduced at a coarse-grain level, e.g., by powering or switching off groups of cells at a higher abstraction layer. By lowering the supply voltage to V_{dd_L}, which equals $1\,V$, we can further reduce leakage power consumption. Experimental results have shown that data retention will still be ensured at supply voltages down to $400mV$. A combination of all three power saving mechanisms in a 6T SRAM cell is shown in Figure 5.29.

SRAM cell	∅ P (nW)	max{P} (µW)	min{P} (pW)
LP 4T hvt	232.9	21.27	49.59
5T hvt	321.4	22.1	181.4
6T hvt	458.7	44.67	166.1
7T hvt	368.3	26.53	167

Table 5.9: Simulation results with *hvt* transistors, stacking and DVS

In order to achieve an average power consumption of $232.9nW$ at a clock requency of $200MHz$ and full data retention like shown in Table 5.9, we combined all three power saving methods introduced in the chapters before with careful transistor sizing of an efficient memory cell design. The newly developed low-power (LP) 4T SRAM cell is presented by Figure 5.30.

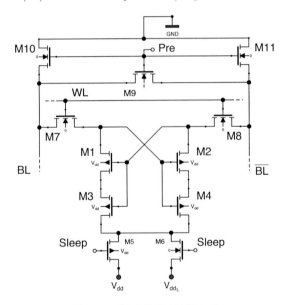

Figure 5.30: LP 4T SRAM cell

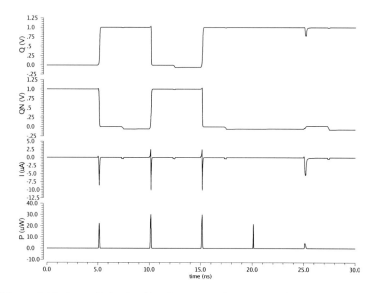

Figure 5.31: Power dissipation P and power supply current I of a LP 4T SRAM cell

The simulation was done by injecting a $HIGH \rightarrow LOW \rightarrow HIGH$ sequence and one read cycle at the end of the simulation time, which can be seen in Figure 5.31. By comparing the results of Figure 5.31 with the outputs shown in Figure 5.23, we see a reduction in both, power and current spikes. Looking back on the continuous improvements added to each cell type, we see the benefits in reduction of average power consumption in Figure 5.32. In addition to that, the minimum and the maximum consumed power is displayed in Figure A.4 and in Figure A.5.

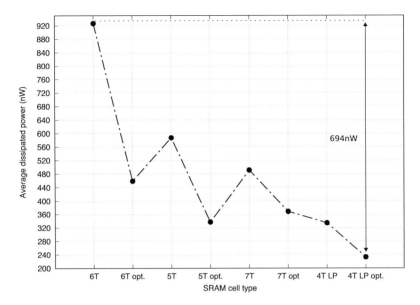

Figure 5.32: Power dissipation reduction

Further information about the improved maximum and minimum power dissipation are shown by Figure A.4 and Figure A.5. These illustrations and Figure 5.32 depict the advantages of combining an SRAM cell design with inherent power efficiency and appropriate modifications for even better energy savings in applications with limited resources. In order to complete the whole evaluation of the newly implemented design, it is also important to involve the SNM and WNM results at this point. Both results are shown in Figure 5.33 and 5.34 respectively.

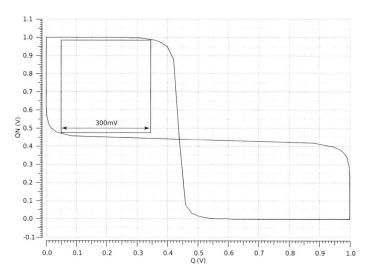

Figure 5.33: SNM evaluation of the LP 4T SRAM cell

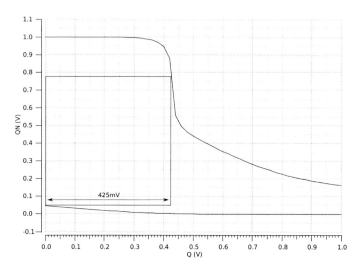

Figure 5.34: WNM evaluation of the LP 4T SRAM cell

So by referring back to Table 5.2 the static and write noise margin results of the LP 4T SRAM cell are quite considerable and therefore added as additional column in an extended Table 5.10 for the sake of a better comparison. These results show that the LP 4T SRAM outperforms the other

71

cells in terms of SNM and shows third best results when it comes to WNM, which is an acceptable result.

SRAM type	4T	4T Noda	5T	6T	7T	LP 4T
SNM (mV)	92	155	250	280	270	300
WNM (mV)	500	430	300	425	415	425

Table 5.10: SNM and WNM comparison of all cells

All SNM and WNM results are additionally displayed in Figure 5.35 and graphically compared to each other.

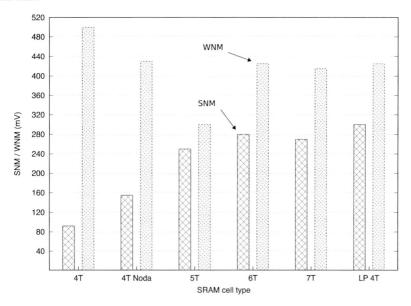

Figure 5.35: SNM and SNM comparison

5.3 Timing considerations

Despite the fact that timing aspects play a minor role for volatile memory cells used for configuring LUTs, a closer investigation of, e.g., the maximum operating frequency f_{max} is helpful to sound the limits of these circuits for their intended usage. In special cases like critical real-time calculations, fast reconfigurability of a programmable logic device may be a inevitable requirement. This

72

maximum operating frequency can be determined by Equation 5.4:

$$f_{max} = \frac{1}{t_{HL} + t_{LH}} \tag{5.4}$$

The summands t_{HL} and t_{LH} display the time necessary for a $HIGH \rightarrow LOW$ and $LOW \rightarrow HIGH$ transition respectively. All considered SRAM cells have been investigated upon these characteristics and compared against each other. The correspondent results are summarized in Table 5.11. The results reveal the superior performance of the LP 4T SRAM cell in terms of elapsed time for both t_{HL} and t_{LH}. In direct comparison to the reference 6T SRAM cell, we achieve an improvement of approximately 60% for the $LOW \rightarrow HIGH$ transition. The improvement for the complementary operation $HIGH \rightarrow LOW$ is less, but time savings of approximately 36% are still noteworthy.

SRAM cell	t_{LH} (ps)	t_{HL} (ps)	f_{max} (GHz)
LP 4T	40.22	38.74	12.66
5T hvt	58.24	132.41	5.25
6T hvt	102.53	60.7	6.12
7T hvt	62.47	380.87	2.25

Table 5.11: Transition times and maximum operating frequency

For having a better overview about the achievable slew rates of all related designs, the respective results are visualized in Figure A.3. The LP 4T SRAM cell outperforms the other cells in each considered aspects. The maximum operating frequency gives an impression about the capabilities of this newly developed cell to be used for calculations in critical real time environments. Figure 5.36 highlights the inversely proportional dependency of all designs between the slew rates of t_{LH} and t_{HL} and the maximum operating frequency f_{max}: the smaller these slew rates are, the higher f_{max} will be. Furthermore, the LP 4T SRAM design offers well balanced numbers for both, the rising and falling edge during a transition. This feature allows to use this cell out-of-the-box without any further modifications to the transistor parameters. It should be mentioned, that this design was not optimized for short channel effect suppression. The recent proceedings in process technology lead to a continuous design shrink, which come along with a significantly higher yield in manufacturing. The downside of these achievements are undesirable physical effects, e.g., the short channel effect. The smaller the channel length becomes, the higher leakage currents in standby mode will be, due to tunneling effects of electrons from drain to source even without establishing a steady voltage $V_{gs} > V_{th}$ at the gate of a transistor.

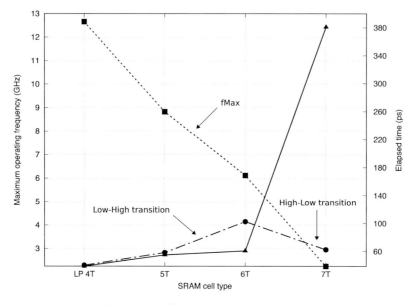

Figure 5.36: Slew rates of t_{LH}, t_{HL} and f_{max}

5.4 Conclusion

Different, existing and published SRAM cell designs were described in Section 5.1. All of them were designed to face potential constraints of applications with limited energy resources and different techniques were applied in each case. Some proposed memory cells use CNFETs, TFETs, memristors or auxiliary circuitry to establish proper functionality and considerable results in terms of power savings. In addition to that, some of these designs presume availability of more than two bitlines as additional control signals are incorporated. Near threshold or sub-threshold operation was used to extend battery runtime while accepting penalties in performance and further modifications against low stability. The proposed memory cell of this work differs from these publications as no extraordinary transistor types are required. In order to ensure a stable operation for both, read and write activities, it was crucial to ensure that all transistors will not work in operating areas where proper switching can not be guaranteed. The typically complementary outputs of dual-ended SRAM cells are an ideal starting point for transformation into dynamic logic, which adds the ability of fast operation at comparably low frequencies. Despite this advantage, the newly proposed design sticks to static logic as its main purpose is to realize configuration memory where rather low leakage currents are of interest than top switching frequencies. Investigation has shown that the choice of a proper memory cell design has a massive impact on power consumption. Regardless of having either static or dynamic power dissipation in focus, if an inherent power saving design is not given, additional modifications will barely compensate any drawbacks emanating from an inappropriate design selection. Based on this initial idea, a dense baseline design consisting of four transistors only, was figured out to be the starting point for further redesigns and improvements. Different modifications have been applied to improve static and dynamic power dissipation, turning the original design into the LP 4T cell. Having a fair comparison in mind, the same low-power improvements have been also added to alternative designs. The simulation results have reconfirmed once more that the LP 4T cell outperforms its competitors and delivers the best overall performance. A further cut-down of power dissipation could be achieved by moving to a more advanced process technology, similar to related research work. Nevertheless, all added improvements require a complex manufacturing process, raising the total costs for the development and release of this specific design. This should be taken into consideration when dealing with tight project budgets.

Chapter 6

Low-power data flip-flop

6.1 Basic considerations

In general, we can distinguish between two different types of storage elements used in registers of, as in processors, latches and FFs. Both designs inhibit their pros and cons and are typically designed to serve different purposes, which are illustrated in Figure 6.1.

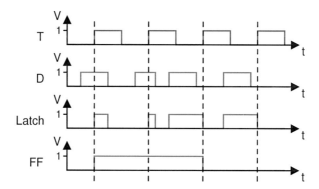

Figure 6.1: Basic working principle of latches and D-FFs

The inputs for both implementations are a clock signal T and an input signal D, which can be a sequence of pre-defined voltage levels based on a randomly generated sequence of input data. A latching circuit shows full transparency once T is set to *HIGH*: regardless what kind of logic value is applied on D, the output of a latch follows every change on its input node immediately. On the other hand, once T is driven to *LOW*, this kind of circuit latches or stores the latest input applied on D before the clock signal is changed from 1 to 0. Depending on the respective applications, this special feature called transparency might be a desirable behavior or not. To overcome this problem and to have a real alternative to latches, FF circuits were invented, which are sensitive to the rising or falling edge of the clock signal. In special cases, even both edges can be used to evaluate the applied data.

Figure 6.1 shows the function of a positive-edge driven FF. As the edge of T rises from 0 to 1, the FF samples the data D and stores it until the next rising edge of T, regardless of all changes on D in between. Hence, the transparency effect which of latching circuits is avoided. Furthermore, an additional evaluation of D could be implemented to sampling the input signal even during the falling edge of T, which leads to faster operation. However, this comes along with some modifications and should be rather decided case by case. Certain design offer the possibility to be configured either to work as latch or as FF, but since the transparency effect is of no benefit in many cases, this paper focuses on the investigation of a low-power FF. D-FFs are an integral part of many designs used in different applications, like storage registers, counters, frequency dividers, etc. FPGAs resort on these circuits in each slice, which is a basic computational element, shown in Figure 6.2.

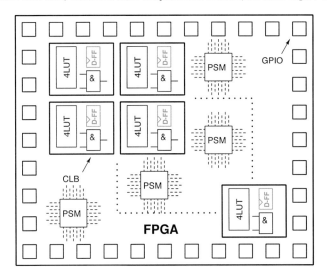

Figure 6.2: Simplified FPGA architecture with highlighted D-FF related blocks

Figure 6.2 depicts the internal structure of a FPGA and highlights in orange color the periodic repetition of D-FFs. Thus, it can be assumed that there is a massive number of D-FFs inside a chip and is worth to be considered for further optimization. Usually, each D-FF is connected to a LUT and some more basic logic gates. So, taking this circuitry and doubling it leads to the realization of a slice. Each slice contains one D-FF for storage of computed values prior to forwarding them to the next configurable logic block (CLB). Since even a low-cost FPGA, e.g. Xilinx Spartan 3A, contains up to 8320 CLBs [105], one can see the strong impact on area and energy consumption of these clocked devices. The relation between consumed power and the supply voltage, load capacitance and system clock can be seen in Equation 6.1.

$$P = \alpha C V_{dd}^2 f_{Clk} \tag{6.1}$$

The activity factor α represents the cadence of write requests, C is the load capacitance and f_{Clk} the operating frequency of a circuit. A reduction of α can be achieved by special memory cell designs [111] or alternatively with auxiliary comparator circuitry. Another efficient approach is reducing the operating voltage. This can be achieved by techniques like dynamic voltage scaling (DVS), which was evaluated in various publications [1]. Power gating is certainly the strongest way to achieve a measurable reduction of energy consumption. However, this can be only applied if there is no focus on data retention. A further possibility for raising the energy efficiency is lowering the clock frequency f_{Clk}. Circuitry, which is not timing critical, can be clocked down to a minimum speed which ensures a reliable operation of the system. If certain circuit parts can be completely stopped while retaining stored logic values, full clock gating can be a feasible solution to save power [112]. Both methods can be combined on a coarse-grain or fine-grain level.

These techniques are only an extract of a set consisting of different methods on how to handle the challenges of demanding functions. A majority of these solutions require additional circuitry to be added and implemented at a higher architectural level. Our approach goes one step further and is based on direct circuit level improvements to a D-FF by reasonable selection of a suitable D-FF cell design and substantial modifications of the internal cell circuitry to achieve better efficiency. The improvements achieved on that level are essential for important energy dissipation suppression and are an inevitable step for optimization to be combined with architectural amendments.

6.2 Selected legacy designs

Different concepts have been introduced in the recent years. Whilst latches are level-sensitive designs, flip-flops are egde-sensitive. Latches are transparent and therefore not suitable for timing-critical applications due to possible glitches in the signal path. For avoiding glitches and in consequence timing problems in complex designs, many flip-flop designs implicate the principle of cascading *master-slave* D-FFs. This standard design in shown in Figure 6.3.

Both, master and slave unit, consist of a feedback loop of inverters and transmission gates. Once *Clk* is set to *HIGH*, the input data provided by D is latched in the master circuit. At this point, the transmission gate (TG) connecting master and slave circuit, is in cut-off mode and therefore avoiding any glitches, e.g. direct throughput of D to Q. When *Clk* is set to *LOW*, the stored data at the output of the master circuit is latched by the subsequent slave unit and provided at the output node Q. Any changes of D do not influence the logic value stored at Q due to the fact

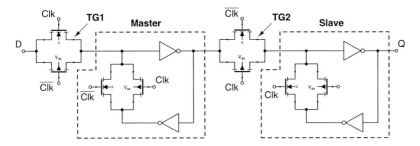

Figure 6.3: Master-slave arrangement

that both transistors of TG 1 are in cut-off mode. This legacy design was the starting point for numerous variations in the past. All simulations have been performed with Cadence tools and a $90nm$ technology provided by TSMC at an ambient temperature of $27°C$. The clock frequency was set to $250MHz$ and the respective simulation results are shown in Figure 6.4.

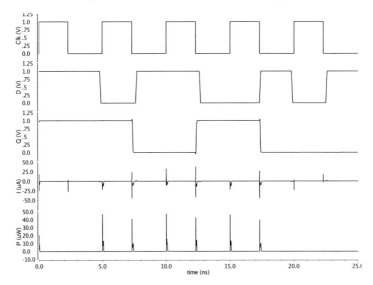

Figure 6.4: Simulation results of Reference D-FF

80

SET D-FF

A simplified implementation is shown in Figure 6.5. Whilst the reference design of a D-FF uses 16 transistors in total, this design consists of 10 transistors only, leading to a higher chip density and reduced manufacturing costs [113].

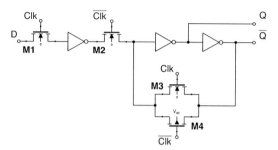

Figure 6.5: Single Edge Triggerd D-FF

Instead of 4 TGs, this design works with 1 TG and achieves the same function by replacing the remaining TGs by nMOS transistors. This reduction of transistors comes along with cutting down the number of slower and larger pMOS transistors. Furthermore, this implementation provides the generation of both Q and $\overline{Q}$. The functionality of the SET D-FF is similar to the reference design: glitching is avoided by complementary control of both pass-transistors $M1$ and $M2$. Latching and generation of the output values is done in the feedback loop after the activation of $M2$. Analog to the previous standard design, this concept relies on the preparation of complementary Clk signals, which requires additional circuitry for signal generation. The simulation results are illustrated in Figure 6.6. The slew rate of Q during a $HIGH \rightarrow LOW$ switching event is noticeable weaker than of its $\overline{Q}$ counterpart. This goes back to the additional inverter right after Q is generated within the signal path and which is used to achieve a higher slew rate of $\overline{Q}$, but also adding a slight delay. To improve the slew rate of Q, an appropriate adaption of the signal chain's second inverter transistor parameters should be done.

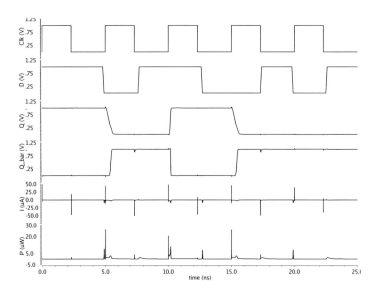

Figure 6.6: Simulation results of SET D-FF

Low-power (LP) D-FF

Another variation, which displays an attempt on how to optimize a D-FF with respect to power consumption, is shown in Figure 6.7.

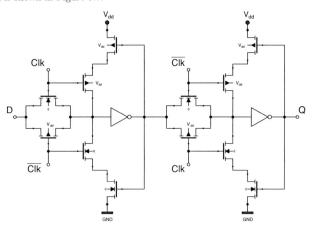

Figure 6.7: Low-power modification of D-FF

82

The key aspect of this design is to eliminate short-circuit power dissipation from the feedback path [114] due to the tristate inverter. Although keeping the same number of transistors like in the reference design, considerable power savings can be achieved.

In direct comparison to the SET D-FF, Figure 6.8 depicts a better slew rate of the output signal Q, regardless of considering a $HIGH \rightarrow LOW$ or $LOW \rightarrow HIGH$ transition. However, this design does not support provision of complementary outputs, which would come along with further modifications.

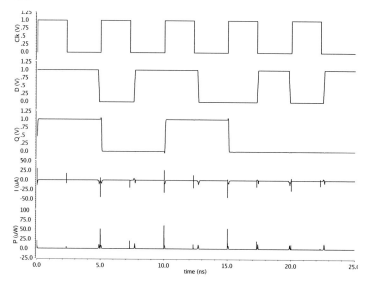

Figure 6.8: Simulation results of Low-power D-FF

PPI D-FF

In order to get a better performance of a conventional D-FF, the Push-Pull-Isolation (PPI) D-FF was presented in [114]. The main advantage of this implementation is the reduced clock-to-output delay from two gates in the reference design to one gate in the PPI D-FF, which is shown in Figure 6.9.

The insertion of an inverter and a TG between the output nodes of master and slave latches provides a push-pull effect at the slave latch. In consequence, the input and output of the inverter in the slave unit will be driven to opposite logic values during operation. This design is approximately 31% faster than the reference D-FF, but has a power overhead of 22%. To counter the increased power consumption two pMOS transistors, $M1$ and $M2$ are added to the feedback loops in the master and slave latches. In direct comparison with the conventional D-FF, the PPI D-FF improves speed

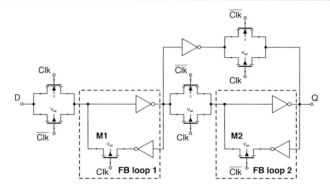

Figure 6.9: Push-Pull-Isolation D-FF

by 56% at an expense of 6% of additional power dissipation. The respective simulation results are shown in Figure 6.10.

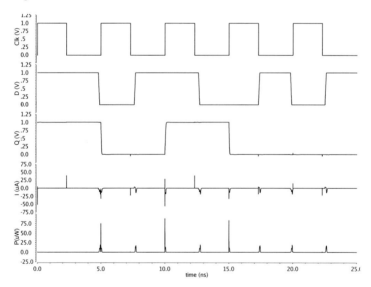

Figure 6.10: Simulation results of PPI D-FF

For achieving comparable results, all designs have been simulated with the same test circuit and same stimuli inputs. The related simulation environment is shown in Figure B.1. The input signals D, Clk and $\overline{Clk}$ are provided by the driver circuitry. Since the signal transition through a simple inverter adds some delay between both Clk signals, additional circuitry for synchronizing these signals must be added. For the sake of simplicity, this is not shown in Figure B.1. The load

84

consists of two additional inverters at the output nodes. The design under test (DUT) is powered by an independent voltage source to enable a precise comparison of the D-FF designs in scope of this research work. For the low power and PPI D-FF, which are not supporting the generation of $\overline{Q}$, the test circuit has been appropriately adapted. For all introduced cell designs in this thesis, the average power consumption, the maximum and minimum power consumption during simulation time were traced and summarized in Table 6.1. These results show that the reference D-FF dissipates the highest average power consumption by $1186nW$, due to lack of power savings measures. A remarkable improvement can be seen by comparing this result between the reference design and the LP D-FF, especially when having a closer look on Figure 6.11. The maximum power dissipation confirms this result by revealing a higher consumption by the factor of approximately four in direct comparison with the optimized low-power D-FF. However, this result was expected and highlights the improvements of previously introduced designs.

D-FF Type	$\varnothing\, P\ (nW)$	$\max\{P\}\ (\mu W)$	$\min\{P\}\ (fW)$
Reference	1186	233.3	51.47
SETD	280.3	26.21	22.39
Low-power	272.7	61.55	19.92
PPI	435.4	88.71	28.01

Table 6.1: Simulation results of dissipated power P

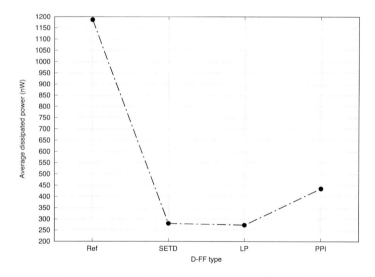

Figure 6.11: Comparison of average power dissipation

On the other hand, similar results are reflected by measuring the leakage current of each design, shown in Table 6.2. The reference D-FF exhibits the highest average power supply current I by $1262nA$, which is approximately fivefold higher than the average power supply current I of the low-power D-FF and is also illustrated in Figure B.2. Analog to the average supply current, the maximum current is also allocated to the reference design and points out that all power-optimized variations perform better in terms of energy efficiency. These results are also highlighted in Figure 6.12.

D-FF Type	$\varnothing\ I\ (nA)$	$\max\{I\}\ (\mu A)$	$\min\{I\}\ (\mu A)$
Reference	1262	346	336.3
SETD	265.7	50.41	48.94
Low-power	235.1	45.9	28.83
PPI	403.7	56.35	39.4

Table 6.2: Simulation results of power supply current I

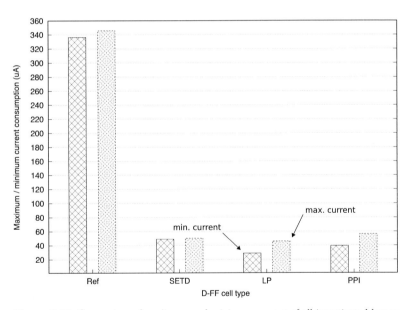

Figure 6.12: Comparison of maximum and minimum current of all investigated legacy designs

The respective simulation results are shown in Figure 6.13, which illustrates the input signal D, the clock signal Clk and the respective power dissipation output profiles for the presented input sequence with an alternating $LOW \rightarrow HIGH \rightarrow LOW \rightarrow HIGH$ sequence.

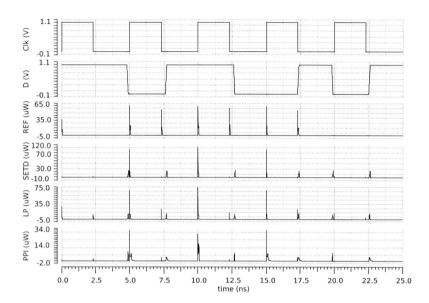

Figure 6.13: Comparison of dissipated power

All designs exhibit strongly varying power consumption for each transition on the input nodes during the rising edge of the *Clk* signal, which comes along with an exploitable vulnerability for side-channel attacks. Glitches can be identified during the falling edge of *Clk*, which indicates weaknesses in the latching mechanism of master and slave latch, therefore revealing undesired transparency. Beside the analyzed designs further research work was done on D-FFs. These designs were not analyzed in detail as they depict variants of the closer investigated designs or exhibit features, which highlight a different application focus. Nevertheless, for the sake of completeness, these designs shall also be briefly described here. For achieving a reduction of leakage power with regard to the reference design, two inverters were replaced by sleepy stack inverters [66]. These special inverters introduce an active and *sleep* mode to the D-FF design and are used to save power and to retain data in *sleep* mode. This feature can be added to further FlipFlops for reducing power consumption in a considerable way. The negative aspect here is the increased number of transistors, which is 34 after implementing the sleepy stack inverters. A different D-FF circuit implements positive feedback source coupled logic (PFSCL) and is based on triple tail logic [67]. By this combination, decent power savings could be made evident, while revealing disadvantages in performance and number of required transistors (32 in total). Further research work focused on further simplification of the reference D-FF and on adding self voltage level (SVL) control [68]. The SVL control is added between the core D-FF cell and V_{dd} and *GND*. In dependance of the *Clk* signal, the D-FF cell is either connected to V_{dd} and *GND* or decoupled from them and leading to virtual V_{dd} and virtual GND nodes. The core D-FF cell is reduced to five transistors in total and reduces consumed power, however, this reduction leads to transparency from input to output when *Clk* is *HIGH*. This effect is not desirable and should be avoided. Different variations of the true-single-phase clock (TSPC) D-FF were published as results of research activities [69] and a low-power optimized variant was also presented. None of the previously presented designs are optimized in terms of static leakage current suppression or energy recovery during runtime which will be key aspects of the presented design in the next section.

6.3 New power saving modifications

Storing and processing logic values in flip-flops, registers and memories leads to charging and discharging of parasitic capacitances, which are an essential part of each integrated circuit. The development of a Sense Amplifier Based Logic (SABL) D-FF was an intermediate step towards the development of the charge recycling (CR) D-FF. The SABL D-FF inhibits a combination of differential and dynamic logic with the goal to merge the advantages of both principles. For a better understanding of these principles, they are drawn as block diagrams in Figure 6.14. In general, it can be stated that differential logic adds some penalty in terms of additional wiring inside the chip due to the necessity of complementary inputs and outputs. So it is as long a penalty if a designer does not exploit the advantage of increased speed and the benefit of avoiding full-swing signal propagation. As only the difference between complementary signals like In and $\overline{In}$ needs to be sensed, differential circuits operate usually faster than their single-ended counterparts. Here, designing dedicated sense amplifiers come into play and must be handled with care, as sense amplifiers offer many possibilities to be designed for various application goals. Nevertheless it should be also mentioned, that differential circuits offer more robustness against noise impact on the signal lanes.

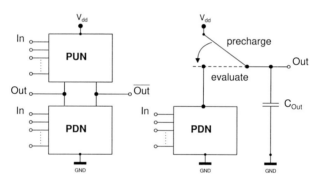

Figure 6.14: Principles of differential (left) and dynamic logic

Developing a SABL D-FF leads to the inevitable fact that a SABL inverter needs to be developed first as the SABL D-FF follows the same work principle like all Master-Slave-FlipFlop architectures (Figure 6.15).

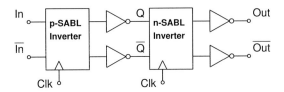

Figure 6.15: SABL Master Slave Arrangement

The SABL D-FF is categorized as dynamic logic, therefore it is not allowed to connect SABL gates to each other without using Domino or np-logic. In this case, the decision was taken to use np-logic, therefore two different SABL inverters (n-type and p-type) have been developed and their correct function was verified by appropriate test runs. Figure 6.16 shows the p-SABL inverter, which works a little bit different than its n-type counterpart.

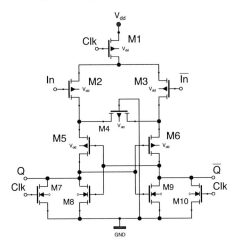

Figure 6.16: p-SABL inverter

As dynamic logic goes back to the *precharge & evaluate* mechanism, each gate precharges its outputs to a certain voltage level. In contrast to the n-type SABL inverter, the p-type SABL inverter 'precharges' Q and $\overline{Q}$ to *GND* instead of V_{dd}. In the following *evaluate* phase one of the output nodes will be charged to a voltage level to approximately V_{dd} while the other one remains at *GND*. The architecture of the SABL D-FF illustrated in Figure 6.15 works as follows: The p-SABL inverter reads and stores both inputs *In* and $\overline{In}$ at the falling edge of *Clk* at Q and $\overline{Q}$. In the next step, during the rising edge of *Clk*, both values are evaluated by the following n-SABL inverter and stored at the output nodes *Out* and $\overline{Out}$. Hence, all input data are stored exactly during one full clock cycle. An example implementation of a n-SABL inverter is shown in Figure 6.17 and the respective simulation results in Figure 6.18. The simulation curves show the correct functional behavior of this inverter and its special characteristic during operation: alternating precharge (*Clk LOW*) and evaluation (*Clk HIGH*) phases. One essential benefit of this design is the almost equal power dissipation during both phases, which adds essential value to countering DPA attacks. This can be seen by evaluation of the current spikes in Figure 6.18.

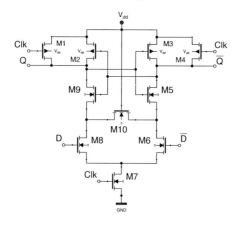

Figure 6.17: n-SABL Inverter

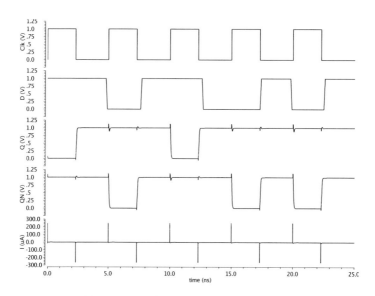

Figure 6.18: SABL Inverter simulation results

91

Figure B.4 shows the SABL MS D-FF at circuit level and the respective simulation results are shown in Figure 6.19. The results illustrate that the SABL MS D-FF works as desired and stores the input value within 1 clock cycle. Q and $\overline{Q}$ display the output voltages of the p-SABL inverter, which acts as the *master* latch. However, designing a D-FF by using dynamic and differential gates and connecting them by np logic results in a total amount of 28 transistors, which is a quite high number compared to the introduced legacy designs.

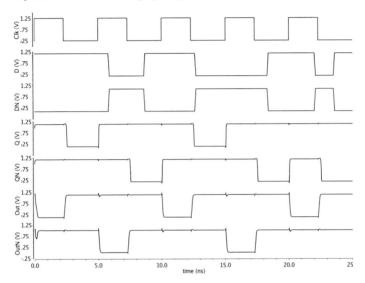

Figure 6.19: SABL MS D-FF simulations results

Since the CR D-FF also features dynamic logic, periodic charge and discharge cycles are an integral part of the intended function and require special attention during the design. Similar to the introduced SABL inverter this design works with two alternating phases during runtime: precharge and evaluate, which are both triggered by the *Clk* signal. Whilst *Clk* turns to *LOW*, M5 is turned on and in consequence also switching on the pMOS transistors $M3$ and $M6$. Illustrating a critical point with respect to power savings within an integrated circuit, the precharge phase is the more deciding one. Due to the fact that these transistors are therefore in a conducting state, the capacitances at the output nodes Out and $\overline{Out}$ are shortened. Hence, not discharged electrons at one of the complementary output nodes are used for charging the previously discharged output node. This effect is used for equilibrating electron charges and thus relieving the battery due to the fact that less energy is needed. This is a strong method for achieving a better performance in terms of dissipation reduction during dynamic behavior. After *Clk* applies a logic *HIGH* at the gate of $M1$, this transistor is turned off whereas $M5$ is turned on and subsequently starting the evaluation phase in terms of sensing the difference between the complementary inputs D and $\overline{D}$.

One of the various benefits of sense amplifier based logic is that even a small voltage difference between both input signals will be sensed and evaluated, providing a higher speed of the D-FF.

6.3.1 Dual threshold CMOS

Leakage currents I_{leak} in D-FFs during standby contribute to a significant amount of total dissipation loss. As D-FFs in FPGAs can be bypassed if their function is not required and therefore put into standby mode, this undesirable leakage effect can not be neglected. By adding dedicated countermeasures, appreciable power savings can be achieved without investing much effort for realization. This can be realized by the usage of transistors with a high threshold voltage V_{th}. Transistors with a high V_{th} require a proportional higher V_{gs} voltage at their gate nodes in order to be turned on, which implies a mitigation of leakage currents. This method can be combined be applying a negative V_{gs} for leading transistors into a deep turn-off status and therefore supporting suppression of leakage currents. This technique should be only applied carefully on circuit parts, which are not timing-critical since higher threshold voltages usually equal in slower signal transition. All transistors in our design are high V_{th} transistors for the sake of strongest suppression of leakage currents.

6.3.2 Multi-oxide technology

Closely related to the previous section, static power dissipation can be further decreased by improving the tunneling-barrier for electrons. Undesired tunneling of electrons through the gate to bulk, leads to current flows which shall be eliminated. The relation between I_{leak} and the tunneling-barrier is shown in Equation 6.2:

$$I_{leak} \propto A(\frac{V_{ox}}{T_{ox}})^2 \qquad (6.2)$$

Increasing the tunneling-barrier can be realized by increasing the gate oxide thickness T_{ox}. A higher oxide thickness leads immediately to a reduction of the tunneling current density I_{leak}, following the goal to extend battery lifetime of mobile devices even in standby mode. The drawback of this technique is similar to the previous one: penalty of the circuit speed may occur if not applied carefully. Based on this reason, we decided to use high T_{ox} transistors for $M1$, $M5$ and $M10$. All of these transistors are not timing-critical, since $M4$ is used to activate a dedicated sleep mode and $M5$ for balancing the outputs. All of these functions are not slowing the circuit speed.

6.3.3 Clk- and power-gating

For further reduction of dynamic power dissipation, cutting off the Clk signal leads to transfer the circuit to a hold state, while maintaining the stored data inside the latches. Circuitry, which is not executing different operations over runtime, can be kept in a *WAIT* state, ready to continue calculation whenever the Clk signal is set to $HIGH$ again. In the proposed design, $M5$ & $M10$ are used for stopping the D-FF from operating, but still keeping the correct data at the outputs of the cross-coupled inverters. Of course, additional circuitry driving and distributing the Clk signal over a whole design is an indispensable requirement. This can be provided by DCMs, which are not covered by this thesis. In case that data storage is not necessary, gating of the supply voltage is an effective method of how to save power in unused parts of a circuit. Power gating can be applied on different hierarchical levels. The decision made was to follow a fine-grain approach, leading to equipping the proposed D-FF with a power gating transistor $M1$. If the *Sleep* signal turns from 0 to 1, $M1$ is off and therefore disconnecting the D-FF from V_{dd}. When this technique is applied in accordance with clock gating, total rail-to-rail-decoupling (V_{dd} and GND) can be realized.

6.3.4 Stacking

The principle of transistor stacking was introduced in Section 5.2.2. The proposed D-FF features stacking as a design principle, e.g. in the pull-down-networks of the slave latch, realized by $M16$ $M17$ and $M20$ and $M21$.

6.4 Simulation results

As a starting point for further considerations and a better comparison, a CR inverter was implemented as shown in Figure 6.20. The total number of used transistors for this inverter's design is 9 and therefore 1 transistor less than compared to the SABL implementation. Figure 6.21 shows the related simulation results. The benefits of applied charge recycling mechanisms can be clearly seen by the output curve of Q. During each precharge phase, the output nodes Q and $\overline{Q}$ are not charged to V_{dd} but to significantly lower voltage of approximately $660mV$. This voltage is created after the equalizing effect of electron charges is balanced out between both outputs. Without any negative affection of the targeted voltage values during the evaluation phase (full swing range from $0V$ to V_{dd}), charge recycling leads to power savings of approximately 34%, which is an estimable number. As a consequence, this power saving mechanism was marked as an interesting feature to be implemented into the CR-DFF. However, while doing this investigations about CR gates, a further important aspect came up: the constraints when cascading CR logic. In opposite to SABL gates, CR gates are not allowed to be connected to each other by using np-logic. All output nodes of a p-CR gate are not 'precharged' to GND but to the threshold voltage V_{th} instead. Similar to that n-CR gates are not fully 'precharged' to V_{dd} but to a voltage of $V_{dd} - V_{th}$. So the circumstance of not fully discharged output nodes of the p-CR gate might lead to high leakage currents in the following n-CR gate, as exactly this output V_{th} voltage adds the undesired possibility to accidental switched on transistors. The stored electric charge at the outputs of the following n-CR gate could discharge and the cross-coupled inverters within this gate might flip.

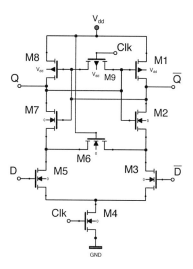

Figure 6.20: CR Inverter

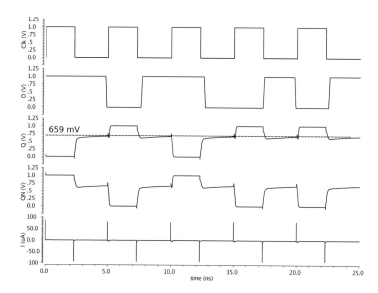

Figure 6.21: Simulation results CR Inverter

To overcome the identified problem, substantial design changes have been made for eliminating the hazard of random data flips within the cross-coupled inverters shown in Figure 6.22.

95

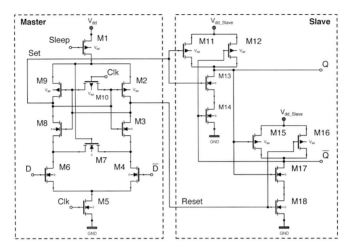

Figure 6.22: CR D-FF

Basically said, the newly implemented CR D-FF consists of a n-type CR inverter and a NAND gate based cross-coupled Set-Reset (SR) FF. As shown in Figure 6.22, the *Master-Slave* principle was still maintained as the CR D-FF serves as *Master* and the cross-coupled NAND gates as *Slave*, which are used to ensure that the differential outputs switch only once per clock cycle. At first, $M10$ is turned on while *Clk* is *LOW* and is used to balance the output nodes *Set* and *Reset*. At this point of time, $M5$ is turned off to prevent any inputs on D and $\overline{D}$ that might impact the electron charge at the outputs of the *Master* circuit. As *Set* and *Reset* are set during the precharge phase to *HIGH* respective V_{dd}-V_{th}, the following cross-coupled NAND gates (*Slave*) maintain the voltage levels stored on Q and $\overline{Q}$, according to the truth table shown in Table 6.3:

Set	Reset	Q_n	$\overline{Q}_n$	State
0	0	1	1	Not allowed
0	1	1	0	Set
1	0	0	1	Reset
1	1	Q_{n-1}	$\overline{Q}_{n-1}$	Store

Table 6.3: Truth Table for NAND SR FF

So this mechanism is responsible for storage of the input data for one full clock cycle. The equalizing transistor $M7$ is continuously driven by V_{dd} at its gate connector and therefore also continuously powered on. The reason for this always-on status of $M7$ is that this MOSFET is used to discharge each internal node during the evaluation phase, so that no remaining internal charges could lead to an unintended bit flip inside the cross-coupled inverters. $M1$ works as power-gating transistor to shut down the *Master* circuit if its function is no longer needed, which depends on an appropriate

96

control of the *Sleep* signal. Adding a dedicated gating transistor to *GND* is not necessary as this design already features decoupling from *GND* from scratch. Nevertheless, power-gating could be applied to both NAND gates within the *Slave* section if more energy savings are necessary, but the typical trade-off between reduction of power dissipation and penalty in terms of speed and consumption will probably come into consideration. It should be also mentioned that the CR D-FF offers a considerable chance for transistor stacking, especially in the PDNs of the *Slave* section. By adding more transistors in series to $M13$ and $M14$ respectively $M17$ and $M18$, a reduction of the standby leakage current can be achieved. In summary it can be stated, that this design offers a certain degree of scalability to push power savings to the limit, but restrictions in terms of desired operating frequency and yield for manufacture should be considered. The CR D-FF senses the inputs D and $\overline{D}$ at the positive edge of *Clk* and stores these data independently from any changes at the input nodes of this circuit. Due to all implemented circuit improvements, an average static leakage current of $173nA$ is achieved, which is sufficiently low to be accepted. During the negative edge of *Clk*, the CR D-FF turns into the precharge phase, where all internal and external nodes are charged. The characteristic curves in Figure 6.23 show one beneficial feature of the CR D-FF over the other discussed designs. This can be seen in the output curve of Q.

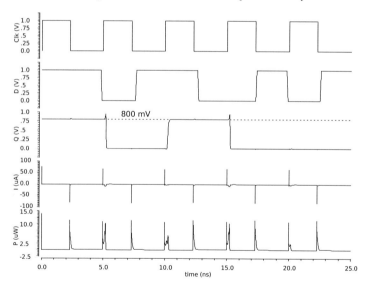

Figure 6.23: Simulation Results CR D-FF

The same simulation was applied with an implemented SABL D-FF, which is shown in Figure 6.24. This circuit lacks the energy recycling feature, which is implemented in the CR D-FF.

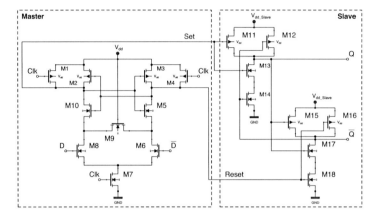

Figure 6.24: SABL D-FF

The respective simulation results are also shown in Figure 6.25. Measuring the average power dissipation led to a result of $442.7nW$. A maximum power dissipation of $21.49uW$ and a minimum power dissipation of $22.73fW$ highlights the competitive results, but which could be improved, especially when discussing about the average result. The maximum difference in consumed power during a switching event is approximately 26.17%, which is the second best result.

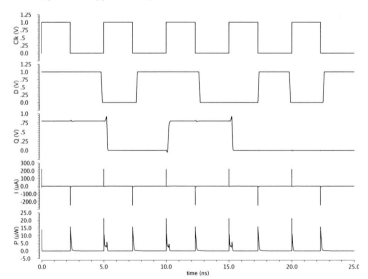

Figure 6.25: Simulation Results SABL D-FF

98

Since this design features charge recycling, the output nodes and all internal nodes are precharged to $V_{dd} - V_{th}$ only, which is beneficial for the energy balance of this circuit. The reason for this is that precharge is finished by achieving an output voltage, which is one threshold voltage below V_{dd}. Thus, the less energy from the power supply is required for precharging the CR D-FF, the more suitable circuitry for low-power applications will be. Based on the reduced voltage range at the outputs of the *Master* latch, it is possible to decrease permanently the supply voltage $V_{dd\ Slave}$. Hence, we choose a supply voltage of $800mV$ for the conventional slave circuit, which supports further power dissipation reduction. For a better comparison, we enhance Table 6.1 with relevant simluation results of the CR D-FF, shown in Table 6.4.

D-FF Type	∅ P (nW)	max$\{P\}$ (μW)	min$\{P\}$ (pW)
Reference	1186	233.3	51.47
SETD	374.1	32.01	22.39
Low-power	275.7	73.89	19.92
PPI	435.4	110.5	172.3
SABL	442.7	21.49	22.73
CR	**303.5**	**13.84**	**27.59**

Table 6.4: Comparison of dissipated power P of all D-FFs

The results in Table 6.4 show that the introduced CR D-FF outperforms most of the previously analyzed designs in terms of average power consumption, which is also illustrated in Figure 6.26.

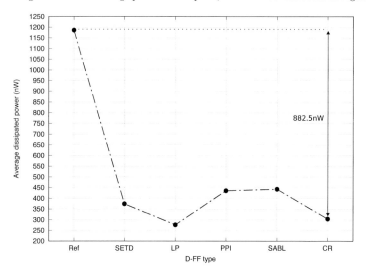

Figure 6.26: Comparison of average power dissipation of all investigated designs

It achieves the second-best performance for average power consumption ($319.7nW$) and the best result for maximum power dissipation ($13.84uW$). The minimum power consumption of $27.59fW$ can be neglected, since the influence of these contributions is not significant for the overall performance of all discussed designs. Even though the conventional low-power flip-flop achieves a slightly lower average power consumption than the CR D-FF, the peak power dissipation is approximately quintuple higher and it offers no resistance features against DPA. Figure 6.27 displays a more detailed view on the power variation from both dynamic D-FFs, whereas Figure 6.28 highlights an overall comparison of the average power consumption of all discussed designs. Additionally, the differences in average power consumption are highlighted in Figure B.8.

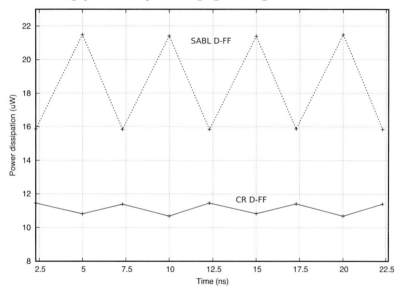

Figure 6.27: Comparison of Average Power Dissipation between SABL and CR D-FF

Even though Figure 6.28 indicates rather small differences in the variation of power consumption coming from SABL D-FF and CR D-FF, Figure 6.27 clearly highlights the reduced difference in power dissipation between both designs. Whilst the consumed energy between each transition of the SABL D-FF varies at a maximum difference of approximately $5\mu W$, the CR D-FF performs better. Here, the variations do not exceed a maximum voltage of approximately $500nV$. It can be clearly seen in Table 6.4 that the CR D-FF provides the most constant power consumption among all considered designs, therefore also providing the best opportunities to be chosen in security-sensitive applications. The smaller the differences in energy consumption between each data transition are, the more difficult a differential power analysis will be, which is always the starting point for a side-channel attack. Hence, the introduced CR D-FF provides both, remarkable low-power

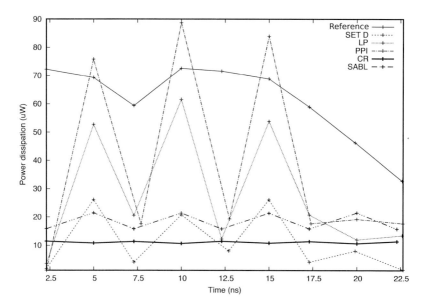

Figure 6.28: Comparison of Average Power Dissipation

characteristics for mobile, embedded circuitry, which comes along with a necessity for robustness against intended attacks. However, benefits in superior energy efficiency and noticeable robustness against differential power analysis come at the cost of a higher number of transistors, shown in Table 6.5 and in Figure B.6.

D-FF Type	Reference	SETD	LP	PPI	SABL	CR
No. of transistors	16	10	16	18	18	**21**
max$\{P\}$ Δ (%)	18.78	94.7	94.03	98.62	26.17	**6.8**

Table 6.5: Transistor count and power variation

This fact usually leads to a penalty in required area for manufacturing, which is certainly an aspect to be considered. A CR D-FF consists of 21 transistors and requires preparation of complementary input signals, which depend on additional wiring and therefore lead to extra area on the chip. On the other hand, this implementation provides also 2 complementary outputs with no delay between both signals and no necessity of additional circuitry for generation. Table 6.5 also emphasizes the differences between the analyzed cells in switching behavior. Whilst the difference of dissipated power of the CR D-FF never exceeds variations of 6.8% in maximum, the results of the alternative designs show much higher noticeable differences. Despite the fact that all designs have been analyzed without putting a stronger focus on speed and timing aspects, further mea-

101

surements on the maximum operating frequency have been done. For this purpose, the elapsed time for each switching transition was measured and compared against each other. Based on these simulation results, the consumed time for a $HIGH \rightarrow LOW$ and a $LOW \rightarrow HIGH$ transition has been measured and summarized in Table 6.6. For the sake of a better overview, these results are additionally illustrated in Figure B.5. The maximum achievable switching frequency f_{max}, which is illustrated in Figure B.7 as a comparative overview, reveals the penalty in operating speed of the CR D-FF, due to the increased number of transistors. However, a maximum switching frequency of approximately $6.4 GHz$ is still a notable result. It shall be mentioned that even better results in terms of speed could be achieved by a further fine tuning of the transistor parameters. Especially p-channel transistors may be a bottleneck when it comes to circuit speed optimization as they are slower than their n-channel counterparts. Thus, keeping the gate length as short as possible whilst maximizing out the gate width would turn into a faster switching behavior. Both, n-channel and p-channel transistors can be modified in this way to get better results as a higher width provides the possibility for higher currents and therefore to faster charging of load capacitances is realized. If circuit speed has priority over other circuit characteristics, this is a reliable method to achieve shorter execution times. This was not the case during the research work done for this thesis as low-power optimization was of first priority.

D-FF Type	t_{HL} (ps)	t_{LH} (ps)	f_{max} (GHz)
Reference	42.5	58.3	9.9
SETD	422	101	1.9
Low-power	43.63	51.58	10
PPI	60.48	79.16	7.1
SABL	56.19	103.21	6.2
CR	**41**	**114**	**6.4**

Table 6.6: Timing comparison of all investigated and newly developed D-FFs

6.5 Conclusion

A number of different and known D-FF designs was depicted and analyzed upon their capabilities to save power in active and standby mode. This was the aspect of highest priority during all investigations. Second, further topics like transistor count and operating speed, have been also addressed and evaluated. Like shown in Table 6.4, the newly designed CR D-FF is second best for average power consumption, which is a fairly considerable achievement. Of course, beating the Low-Power D-FF would still be seen as a desirable goal, but exploring new design extensions come very often at cost of other characteristics. Nevertheless, the CR D-FF shows good results when moving from average considerations to the peak power, which is consumed during transitions. High reliability in safety-critical applications puts additional requirements for the 'robustness' of a circuit in order to harden it against high voltage spikes on the power lanes. At this place, the new design becomes top notch in comparison to the other circuits. The less maximum power is consumed, the less additional slack must to added to design aspects like over voltage sizing. Security issues are becoming more and more important, therefore countermeasures down to chip- / gate-level have to be implemented, e.g., S(ubstitution)-Boxes [115]. Legacy designs lack of appropriate design modifications to raise inherent robustness against DPAs, as their power dissipation during transitions reveals strong variations. Both, the SABL and CR D-FF, mitigate this weakness by minimizing these variations through a better balance in charging and discharging parasitic capacitances, which is highlighted in Figure 6.28 and Figure B.8. The CR D-FF is even better in terms of balanced power dissipation and represents the top notch of all compared circuits by cutting down the power variation to 6.8% only.

Chapter 7

Low power tristate buffer

7.1 Basic considerations

GPIOs are used in almost every integrated circuit as interfaces for communicating with peripheral circuitry. These elements are designed for receiving data as inputs and to output data to other connected devices. Therefore tristate buffers are bidirectional circuits with the ability to receive and transmit logic signals by the same input/output pin. Due to this important function, GPIOs play a major role in the consumed area of a chip and the power consumption in each complex design [105]. Similar to Figure 6.2 in Section 6.1, Figure 7.1 illustrates a simplified block diagram of a FPGA without any additional hard processing cores, but with highlighted GPIO units in orange color.

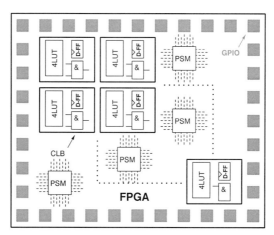

Figure 7.1: Simplified FPGA architecture with highlighted GPIOs

As illustrated in Figure 7.1, all CLBs of this simplified internal hierarchy are surrounded by GPIO blocks. For the sake of simplicity, the internal structure of these blocks is not further detailed in this figure, but will be picked up again in Figure 7.3. In complex systems, several FPGAs may drive an internal bus for different purposes, e.g., data exchange, leading to potential conflicts when different circuits try to write different logic values to the same bus line.

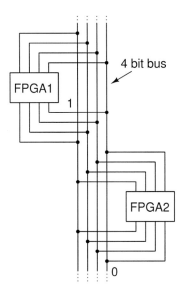

Figure 7.2: Interconnection bus

Figure 7.2 highlights the described conflict and depicts a situation, in which two different FPGAs, connected to the same 4 bit data bus, drive the same line with different values: whilst FPGA1 drives one signal line of the bus with a logic *HIGH* or V_{dd}, FPGA2 tries to do the same but with a logic 0. The consequence is a floating voltage on the interconnection signal line, which is difficult to predict and an undefined state. For this reason tristate buffers play an important role inside each GPIO, since they offer one special output state beside their functionality to pass a logic value from the input to the output node: *highZ*, also called high impedance. Before elaborating any further on the details of this special ability, it makes sense to take a closer look into the simplified, internal structure of a GPIO, shown in Figure 7.3.

A GPIO is divided into three sections: the tristate enabling block at the top of the figure, the output handling and the input handling blocks in the center and at the bottom of the block diagram. The tristate handling block consists of a D-FF and a 2:1 multiplexer, which is used when a registered or non-registered *enable* signal is been applied to the tristate buffer. An additional SRAM cell is used to provide a *select*-signal for the multiplexer and can be fully configured by setting the SRAM cell

106

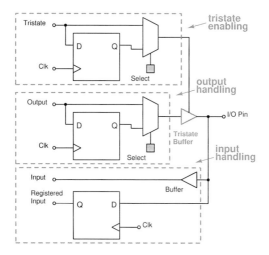

Figure 7.3: Internal structure of a GPIO circuit

as desired. The reason for having a registered tristate signal built-in is usually timing issues, as a D-FF can be also used to delay signal propagation for a dedicated timeframe. Output handling is done in a similar way to the tristate enabling section and drives the input node of the following tristate buffer. The input handling section lacks of the multiplexer and differs slightly from the other blocks as at still provides a D-FF for registered input signals but also exhibits a buffer for voltage level restoration. Therefore, the point of interest here is the tristate buffer, that is being used to connect all logic to peripheral circuitry. In case that the *Tristate* input receives an *enable* signal which is a logic *HIGH* and propagates through the multiplexer, a following tristate buffer cuts off the connection between its input and output node and therefore prevents an undesired throughput from the inputs of the GPIO inside an FPGA to the interconnection bus. Overall, three relevant aspects for optimization in terms of energy efficiency can be identified:

- Subthreshold / standby leakage

- Active power consumption

- *highZ* behavior

The *highZ* attributes of a tristate buffer play an important role due to their ability to decouple this buffer from the remaining signal chain. A careful design of the output transistors inside a tristate buffer offers heavy impact on this ability. It should be stated here, that priority was given on low power characteristics of our newly implemented design. Measurement of the *highZ* state with different output voltages was done after evaluating power consumption of all investigated designs. Furthermore, all measurements were compared with each other to determine which design performs best in general. All considerations about static and dynamic power dissipation, which have been introduced in Chapter 3, are also valid for a tristate buffer. Beyond that, all described power reduction measures mentioned in Section 6.3.1, 6.3.2, 6.3.3 and 6.3.4 also apply here.

7.2 Legacy design

The basic purpose of a buffer circuit is to forward the input value with a certain delay to the output node. Some applications might require the addition of a delay time for synchronizing different data paths. The easiest way to understand the basic function of a buffer is to imagine the logic function of two inverters in series. A tristate buffer adds a third, important feature to this functionality: the $highZ$ state. For a better understanding of the circuit's function, a tristate inverter is shown in Figure 7.4.

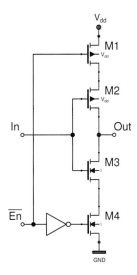

Figure 7.4: Tristate inverter

As long as $\overline{En}$ provides a logic LOW at the respective input node, the transistors $M1$ and $M4$ are turned on and subsequently provide a direct path to the voltage source V_{dd} and GND. As a consequence, the transistors $M2$ and $M3$ work as an inverter and therefore invert all signals applied to In. On the other hand, if $\overline{En}$ turns to $HIGH$, $M1$ and $M4$ are turned off and cut-off the internal transistors $M2$ and $M3$ from the supply voltage and ground path. In this special case, the voltage at the output node Out is floating and undefined. This means that in dependence of this floating voltage, only a very small current will flow either as leakage current from the tristate inverter into the circuitry connected to Out or from the load into the tristate inverter to GND. By adding one additional nMOS and pMOS transistor, the discussed tristate inverter can be modified to a tristate buffer, which is shown in Figure 7.5.

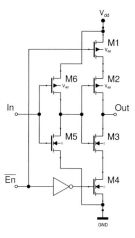

Figure 7.5: Standard design of a tristate buffer

Different aspects of this tristate buffer's behavior have been investigated during simulations by a $90nm$ TSMC (Taiwan Semiconductor Manufacturing Company) technology and a Cadence toolchain (INCISIVE 6.1.5). All simulations, serving the purpose investigating the circuit's dynamic performance, were done at an operating frequency of $200MHz$ and with standard settings for all transistors' dimensions ($120nm$). Since all analyzed designs are not dynamic logic inheriting a dedicated Clk input, the operating frequency was modulated into the switching events of $\overline{En}$. The results of the first simulation run with active inputs are shown in Figure 7.6 and also displayed in Table 7.1 and Table 7.2 . This simulation was followed by further tests for alternative circuit states with the intention to build up a baseline database for follow up comparisons.

Design type	$\varnothing$ P (nW)	max$\{P\}$ (μW)	min$\{P\}$ (pW)
Reference	245	56.75	103.8

Table 7.1: Simulation results of dissipated power P of the reference tristate buffer

Design type	$\varnothing$ I (nA)	max$\{I\}$ (μA)	min$\{I\}$ (nA)
Reference	215.2	230.5	261.4

Table 7.2: Simulation results of power supply current I of the reference tristate buffer

The simulation results shown in Figure 7.6 display the correct function of this tristate buffer, which directly passes the input value to Out whenever $\overline{En}$ is set to LOW. Once $\overline{En}$ applies a logic $HIGH$

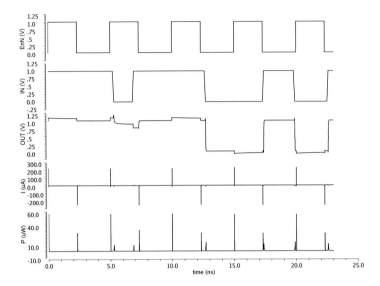

Figure 7.6: Simulation results of dynamic behavior of a standard tristate buffer

to the cutoff transistors, the voltage level at *Out* starts to float and swings between voltage levels above V_{dd} and below $0V$ (*GND*). These floating voltages are not defined and also indicate that the whole circuit is in *highZ* mode. Active power dissipation is of high importance for the estimation of required energy resources, but regardless of these results it is also obligatory to have a closer look on the standby power characteristics when the circuit is led into an idle phase or put completely into standby mode. This means that the data input is inactive and $\overline{En}$ active. The simulation results are shown in Figure 7.7.

For this analysis and for a better observation of the standby current, the simulation runtime was set to $1\mu s$. The simulation curves of both the standby current and the standby power dissipation, show the discharging process of all internal parasitic capacitances after powering on the circuit at the very beginning of the simulation process. Both, standby current and the allocated dissipated power, continuously decrease over time, resulting in an average leakage current of $133.6pA$ and a related average power dissipation of $132.1pW$. The remaining aspect to be considered at this point is the behavior of the reference tristate buffer in *highZ* mode after setting $\overline{En}$ to *HIGH*. As illustrated in Figure 7.2, this special mode is of indispensable importance for a reliable data transmission and communication on a bus, connecting different chips to each other. Figure 7.8 depicts a supposable scenario in a simplified way. Each tristate buffer output node is connected to a signal lane serving as interconnecting bus through the whole system. In dependance of the complexity, there might exist a number of N tristate buffers, each of them being a part of a GPIO. So tristate buffer 1 can be driven by a different input signal than tristate buffer 2 and there might

110

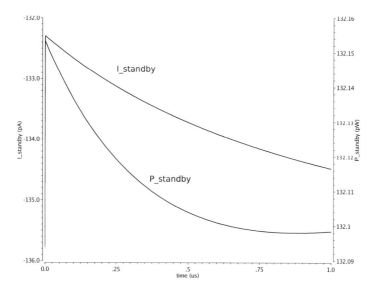

Figure 7.7: Simulation results of idle standby results

be a chance that both of them would try to put these data on the interconnecting signal lane to transmit it while tristate buffer $N + 1$ is trying to put a logic 1 on the bus at the same time. In consequence, arbitrary control system must be added for ensuring a stable and controlled data flow / exchange on each bus, but the subsidiary tristate buffer's *highZ* function is not allowed to be cut down and needs to be analyzed.

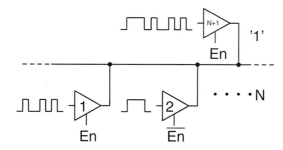

Figure 7.8: Example scenario of various tristate buffers connected to a bus

First of all, it should be stated here that there is no unambiguous answer on this question, since this depends on the voltage which will be applied by the load to the output node *Out* of the tristate buffer. So in dependance of the voltage applied to *Out*, there will be a current flow into an external source / current sink ($V_{Out} = 0V$) or current will drain from the external voltage source into the

111

inverter ($V_{Out} = 1V$). In addition to that, there is always a small throughput from the input node on the output in case that the tristate buffer in $highZ$ is still stimulated with input data, which might be a realistic situation when the related control logic fails. How this scenario may look like is shown in Figure 7.9.

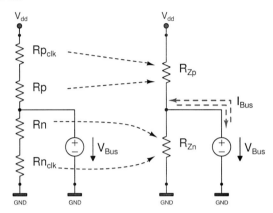

Figure 7.9: Potential impact of bus communication on tristate buffer

By having Figure 7.5 in mind, the transistors $M1$ - $M4$ can also be seen as different resistors Rp_{clk}, Rp, Rn and Rn_{clk}. All resistors in the pull-up and pull-down path can be summarized to equivalent resistors R_{Zp} and R_{Zn}. Voltage source V_{Bus} represents different voltages which will be applied on the output node of the tristate buffer in focus here. Although the most important voltage levels of V_{Bus} will be either $0V$ or $1V$ because considerations during pure digital applications were chosen of most significance, it is still interesting to see how the circuit performs on voltage levels in between. Generally spoken, two different situations, active and inactive inputs, must be considered as they have a significant influence on the resistance of each transistor which is used to couple or decouple the output voltage source from the circuit. This leads to the following cases:

- $In \stackrel{\frown}{=} LOW$

  - $Out \rightarrow Stuck\ at\ LOW$
  - $Out \rightarrow Stuck\ at\ HIGH$

- $In \stackrel{\frown}{=} HIGH$

  - $Out \rightarrow Stuck\ at\ LOW$
  - $Out \rightarrow Stuck\ at\ HIGH$

Based on the assumption that the voltage applied to Out will alternate between $0V$ and $1V$, a DC sweep simulation was done by sweeping the voltage at the output node. The results of both test runs are displayed in Figure 7.10.

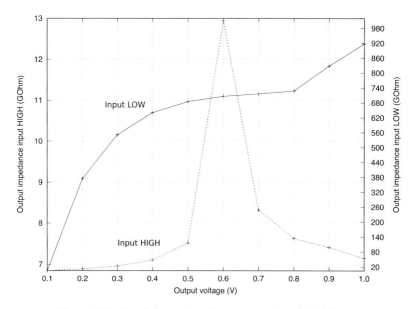

Figure 7.10: Simulation results of output impedance in *highZ* state

Active input data has a remarkable influence on the circuit's capabilities to decouple its internal switching events to the output node. In case of setting the input node to a $0V$ and therefore making it 'inactive', Figure 7.10 reveals a sweetspot in terms of output impedance and which is closely allocated to almost $V_{dd}/2$. Having this striking high output impedance (approximately $1.01T\Omega$) at this voltage range is a desirable situation, since this implements an almost perfect balance between current source and current sink. If input data are applied to *In*, a drop at the output impedance can be observed after simulation. Even with turned off decoupling transistors *M1* and *M4*, the throughput originating from the buffer's input is strong enough to lower the impedance at *Out*. Therefore, a stronger decoupling mechanism would probably lead to better results.

7.3 New low-power approaches

A careful analysis of the reference tristate buffer shown in Figure 7.5 revealed that there is still room left for various improvements. Thus, a noticeable adaption of circuits for sensitive low-power application can only be achieved by a synergy of different power savings measures for imaginable operating states.

7.3.1 Power gating

On our way to develop a low-power tristate buffer, the implementation of a 'hold'-mechanism for standby-phases was an inevitable step. The difficulty here was that the specific design does not comprise clocked inputs which could have been gated like in typical designs built-up by using dynamic logic. Instead of this, a more stringent design technique was applied: power-gating. This modification can be implemented in different ways, by adding a gating transistor between the supply voltage and the circuit or by inserting a transistor between GND and all internal nodes. A third alternative comes along with a combination of both design modifications and is certainly the most effective method in terms of energy dissipation. However, the addition of two transistors to a design with a total number of eight transistors before adding this modification means a transistor count increase of 20% and a high probability for penalty in area footprint. Regardless of the chosen technology for synthesis and allocated continuous proceedings in technology node shrinking, a higher number of transistors is always considered as a drawback. On the other hand, this modification is responsible for a noteworthy limitation of leakage current running through the design under test (DUT), since a full decoupling of the tristate buffer from the supply voltage V_{dd} and GND is realized. In case that a possible penalty in area consumption plays a major role, there is still an option to stick to one of these cut-off techniques mentioned before by accepting certain drawbacks, e.g., leakage currents of not fully discharged internal nodes consisting in each circuit. Power gating shall be triggered by an internal or external signal applied to the gate of the respective transistors. Depending on the architectural details, this triggering voltage can be either provided by dedicated internal control mechanisms or by additional arbitrary control measures at a higher hierarchical layer as described in Chapter 3.1 and illustrated in Figure 3.1. This will be further elaborated in the following chapters.

7.3.2 Leakage current reduction

Weak leakage currents may still come up despite the fact of putting a transistor or circuit into standby mode, as illustrated in Figure 7.11.
These currents are usually in the range of pA and negligible for a single basic circuit like a CMOS inverter, but will certainly grow to a significant factor when discussing their effect in relation to quantity. Thus, special transistors are required to suppress any electron drain from gate to bulk. This problem can be used by appropriate usage of transistors with a higher oxide thickness in

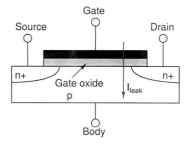

Figure 7.11: Undesired leakage current I_{leak} flowing through gate oxide

comparison to standard transistors. Depending on the chosen process technology, a set of special I/O (Input / Ouptut) transistors may be provided, which could help by handling the leakage current mitigation. I/O transistors are generally used as interface to connect a chip to other electronic devices with higher supply voltages, e.g., $2.5V$ or $3.3V$. In addition to that, this transistor type offers higher oxide thickness (approximately $4nm$) and therefore satisfies the need to handle leakage currents in an appropriate way. Gate oxide thickness T_{ox} should not drop below $2nm$ due to the risk of instant leakage current increase resulting from tunneling effects. The reason behind a relatively thin oxide in core logic is shown in Equation 7.1.

$$C = \frac{\kappa \varepsilon_0 A}{T_{ox}}$$ (7.1)

Here, κ is the relative dielectric constant, ε_0 the permittivity of free space and A the capacitor area. The smaller the T_{ox} value is, then the higher the C value would be, while C value represents the gate capacitance. For performance reasons a high gate capacitance C is certainly a worthwhile goal as it helps to drive current as shown in Equation 7.2.

$$I_{D,Sat} = \mu C \frac{W(V_G - V_{th})^2}{2L}$$ (7.2)

In Equation 7.2, μ is the channel carrier mobility, C is the capacitance density and V_G the applied gate voltage. W and L depict the transistor width and length. Equation 7.2 clearly highlights the trade-off between speed and low power characteristics by designing C and therefore also selecting T_{ox}, have the capability to come up with a crucial push effect towards one of these mentioned directions, which will be more important in depending of the product's planned application. This can be also seen in Equation 3.11, where T_{ox} is in the denominator and therefore has the ability to limit the electrons tunneling through the transistor's gate connector. This results in a reduction of the leakage current in idle / standby state. Despite these benefits it should be mentioned that high T_{ox} transistors have a slower switching frequency than standard T_{ox} have. Hence, adding these transistors should be carefully waived taking a decision upon it. In our case, $M5$ and $M10$ have an increased T_{ox} than the remaining transistors have for keeping the penalty in performance degradation as low as possible. The conclusion of all weighing ups is that this technique is a reasonable extension of previously introduced power gating, provided that these special transistors with a higher oxide thickness T_{ox} are supported by the selected manufacturing technology.

7.3.3 Subthreshold current reduction

Whilst power gating is an effective method for a total shutdown of a circuit when its intended function is not necessary for the ongoing operation of the overall design, there should be an alternative for measurable leakage current reduction when the circuit shall not be totally powered off but still remain in idle state. In addition, even in active operational mode, some transistors will be turned on and others will be turned off, especially in standard CMOS designs like inverters, shown in Figure 7.12.

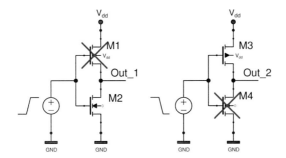

Figure 7.12: Different switching events and operational modes of CMOS inverters

In case that the input source switches from $HIGH \rightarrow LOW$, $M4$ shall be in cut-off mode and keeping the drain connector of $M3$ isolated from GND. Hence, all subthreshold currents through $M4$ shall be eliminated which can be achieved by using high threshold voltage transistor instead of standard threshold voltage transistor. This method might have a negative impact on the maximum operating frequency as a higher V_{gs} voltage needs to be applied. Before inserting high V_{th} MOSFETs a careful analysis of the intended circuit's application cases must be done and their usage should be carefully applied. Nevertheless, these special transistors can not be neglected during the design of low power designs. Based on these considerations the same design was implemented twice: a design build-up by high V_{th} transistors only and one derivation by using these special transistors for decoupling core logic from the supply voltage and GND.

7.3.4   Multi supply voltage

An operating circuit in low power applications should not only be optimized for static power reduction but also for energy efficiency in active mode. As shown in Equation 3.10, the supply voltage has a vast influence on the overall dissipated power. It's obvious that the best approach would be to decrease the global supply voltage V_{dd}, which might lead to the necessity of additional level restorers for a smooth signal transmission to other circuitry. An alternative is the careful partial supply voltage reduction after identifying certain parts of a design, which could be powered by a lower V_{dd}. In case that fast execution time is a leading aspect of development, tuning V_{dd} towards lower levels is not a reasonable way to improve the overall performance. On the other hand, lowering V_{dd} comes along with a slower computation time of the input values, therefore a smaller supply voltage $V_{dd_{low}}$ was only applied to the internal inverter *M6* and *M7*. In principle, there are two different ways how to generate $V_{dd_{low}}$: this can be realized by an external voltage source (illustrated by the additional voltage source $V_{dd_{low}}$ in Figure 7.13) or by exploiting internal voltage nodes (illustrated by the dashed line in Figure 7.13). The second option shows its beauty by an inherent voltage reduction automatism. Once $\overline{En}$ goes *HIGH*, *M10* is turned off and therefore cutting off *M7* from V_{dd}, but keeps the internal inverter still working. Minor adaptions to the width of *M7* have to made due to the decreased internal supply voltage. Nevertheless, both options work well with the low power tristate buffer.

7.4   Low-power tristate buffer

By referring back to the standard tristate buffer shown in Figure 7.5, all previously described measures for achieving better energy savings were stepwise implemented, resulting in the introduction of a low power tristate buffer illustrated by Figure 7.13. As clock gating is no option here, the first modification was the insertion of power gating transistors *M5* and *M10*. These MOSFETs are I/O transistors with the ability to handle supply voltages up to $2.5V$ and a higher gate oxide of $4nm$. Figure 7.5 also depicts the internal tristate transistors *M1* and *M4* which serve the purpose of decoupling both V_{dd} and *GND* from the internal buffer consisting of $M2, M3, M5$ and *M6*. By adding *M5* and *M10* we add a second layer of power gating and the desired function of efficient leakage current elimination in power-off mode.

A simulation was set up to investigate the impact on the higher gate oxide transistors on leakage currents and the simulation results are pictured in Figure 7.14.

By referring back to Figure 7.7, two major differences can be instantly observed. First of all, the transient effects of both circuits appear different due to the high T_{ox} MOSFETs in the modified implementation. In addition to that a comparison of the axis' scales of both figures also depicts the desired effect of additional high T_{ox} transistors. Average leakage current and the related power consumption were analyzed and the results of the reference tristate buffer were opposed to the modified circuit.

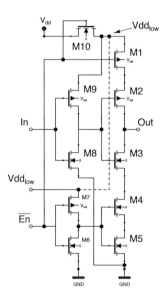

Figure 7.13: Low-power tristate buffer

Measurement	Ref. Design	Mod. Design	Δ%
∅ P (pW)	132.1	17.42	≈ 86.814
∅ I_{leak} (pA)	133.6	1.814	≈ 98.64

Table 7.3: Impact of high T_{ox} MOSFETS

According to all measurements shown in Table 7.3 the desired leakage current reduction can be clearly observed, therefore this modification is considered as a meaning full extension and will be kept for the further evolvement. Further results with an extended simulation time can be seen in Figure C.1. For a further, detailed analysis of applying different power reduction measures, transistors *M1* and *M5* were exchanged against high V_{th} MOSFETs. The idea behind this step is not to reduce leakage currents in off-mode, but also to cut down undesired leakage currents in a powered but idle state. Figure 7.15 highlights once again the internal structure of the newly implemented tristate buffer, in which all standard V_{th} MOSFETS are surrounded by a blue box.

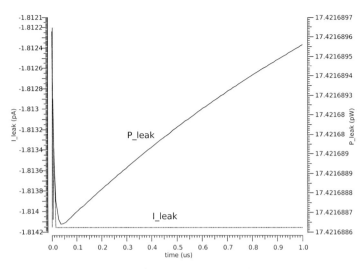

Figure 7.14: Simulation results of power dissipation and leakage current in power-off mode with standard V_{th} transistors as core logic

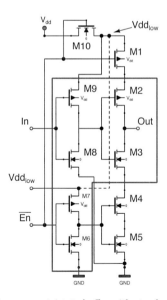

Figure 7.15: Low-power tristate buffer with standard V_{th} core logic

Similar to previous simulations, the simulation time was set to $1\mu s$ for evaluation of the transient response behavior. In was connected to GND, so that no input voltage could impact any internal nodes in steady state / power-off mode. Without a more detailed analysis of the simulation graphs, a comparison of the axis' scales of Figure 7.14 and 7.17 does not reveal a remarkable difference on the first glance. Average I_{leak} equals $1.81pA$ and average correlated power dissipation results in $17.43pW$, which is not a noteworthy improvement when reflecting back on the determined performance written down in Table 7.3. This subsequently led to the decision to replace all remaining standard transistors by their high V_{th} derivatives and the simulations were executed again, with the results as depicted in Figure 7.16.

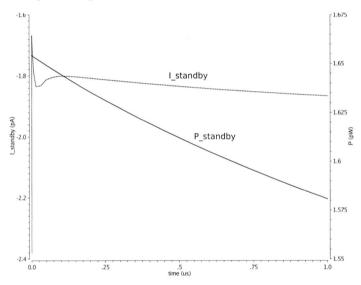

Figure 7.16: Simulation results of power dissipation and leakage current in power-off mode with high T_{ox} power gating transistors

Figure 7.16 reveals a kink shortly after the simulation start due to higher gate oxide thickness which led to a smaller gate capacity and therefore to slower charge / discharge processes driven by a weaker leakage current. The average power dissipation and leakage current were measured and the results are displayed in Table 7.4. As different stages of gate oxides always require additional and expensive steps, a simulation without high T_{ox} transistors was done.

Both results, average I_{leak} and average power dissipation in power-off mode, show definitely a measurable effect by limiting undesired current flow by approximately 55% and related power dissipation by approximately 79%. So it can be clearly seen, that combining both techniques lead to what was intended to achieve when the development was started. At this point, all investigations

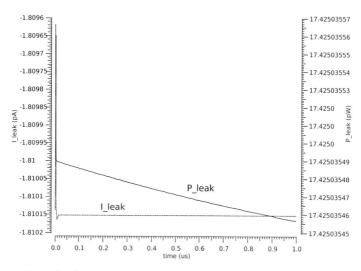

Figure 7.17: Simulation results of power dissipation and leakage current in power-off mode with standard V_{th} transistors as core logic

Design type	$\varnothing$ P (pW)	$\varnothing$ I (pA)
High T_{ox}	4.86	4.67
Standard T_{ox}	0.96	2.1
Improvement (%)	≈ 79	≈ 55

Table 7.4: Impact of high T_{ox} and high V_{th} MOSFETS

and modifications were done for the sake of static low power consumption. However, as the research strategy was to push all improvements to the limit, an additional modification was applied, which is highlighted in Figure 7.18.

A tristate buffer relies on the activation signal $\overline{En}$, which has to be inverted within the circuit (unless it is provided as an external and differential signal) to cut it off from GND and to ensure a stable and reliable tristate mode. This is provided by the internal inverter consisting of $M6$ and $M7$. The idea at this point is to lower the internal supply voltage $V_{dd_{low}}$ to a level, which influences the overall energy consumption in a favorable way but still keeps the inverter operating as intended. The easiest way to integrate this feature is to provide the required inverter supply voltage by an external source, depicted by the solid blue line in Figure 7.18. Auxiliary power management circuits can handle this in a very convenient way, as level shifters are able to generate $V_{dd_{low}}$ just by adding one additional MOSFET which acts as a 'resistor' to generate the needed voltage drop. On the other hand, the question may arise whether $V_{dd_{low}}$ might be also provided

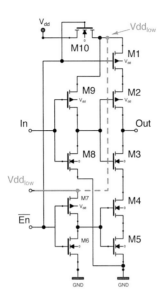

Figure 7.18: Multi supply voltages with different realization approaches

by the tristate buffer itself. This can be achieved by connecting the source connector of $M7$ to the $V_{dd_{low}}$ node, marked by the dashed, blue line in Figure 7.18. The function here is that the internal inverter still keeps working while $\overline{En}$ is a logic 1, driving the whole circuit into $highZ$ mode. Hence, the internally generated $V_{dd_{low}}$ 'supply voltage' shall be sufficiently high to exceed V_{th} of $M4$ and $M5$. This approach works if circuit speed is not negligible at all but does not have the highest priority allocated to it. Necessary adaptions of the power gating transistors' gate length are also indispensable to fine tune internal voltage nodes. As consequence, these steps put higher requirements to the manufacturing process, which may be a questionable fact. Providing the required inverter supply voltage by an external source adds the benefit of adaptiveness for different stages of $V_{dd_{low}}$ and no further constraints to the width or length of any internal transistor. Simulations have shown that the low power tristate buffer works well with $V_{dd_{low}}$ down to $600mV$. However, for having an additional slack which may catch up unexpected voltage drops, $V_{dd_{low}}$ was set to $800mV$ and is therefore 20% lower than V_{dd}. It should be mentioned here that the transistor parameters (W/L) have to be properly set to achieve an optimal switching point at $V_{dd_{low}}/2$ ($\approx 400mV$). The length of both transistors was kept at the minimum size of $120nm$, but the width of $M7$ was set to $260nm$ as p-channel MOSFETs are slower than their n-channel complements and therefore require a wider width. So for the next step, it was necessary to evaluate the effect of this measure during a simulation.

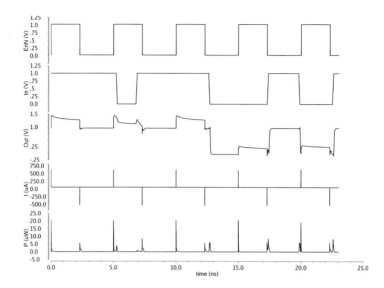

Figure 7.19: Simulation results of dynamic behavior of the low-power tristate buffer

Figure 7.19 shows the dynamic behavior of the low power tristate buffer. The simulation type was a transient analysis, which was set to a runtime of $25ns$ with a random selection of switching events applied to the input of the tristate buffer. For analyzing its behavior at a dedicated operating frequency, a clock of $200MHz$ was modulated on the $\overline{En}$ input, which acts as a switch between buffer and *highZ* mode. Compared to the simulation curves shown in Figure 7.6, it can be seen that the low power tristate buffer is superior in terms of dissipated power during active runtime (see Figure 7.20). By referring to Figure 7.20, a maximal power dissipation of the related simulation results are summarized in Table 7.5 and Table 7.6.

Design type	$\varnothing\ P\ (nW)$	$\max\{P\}\ (\mu W)$	$\min\{P\}\ (pW)$
LP tristate	191.3	29.72	22.36

Table 7.5: Simulation results of dissipated power P

Design type	$\varnothing\ I\ (nA)$	$\max\{I\}\ (\mu A)$	$\min\{I\}\ (nA)$
LP tristate	225.8	194.8	206.9

Table 7.6: Simulation results of power supply current I

Furthermore, an analysis of the standby behavior revealed an improved average leakage current I_{leak} of $24.1pA$ and a related average power dissipation of $22.04pW$. As a final step, the *highZ*

123

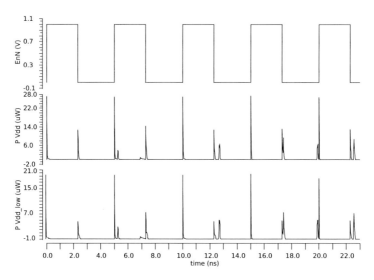

Figure 7.20: Direct comparison of power dissipation with and without scaled $V_{dd_{low}}$

characteristic was investigated for having a better comparison to the reference design. The simulation results are displayed in Figure 7.21 and further results are depicted in Figure C.3 and Figure C.4.

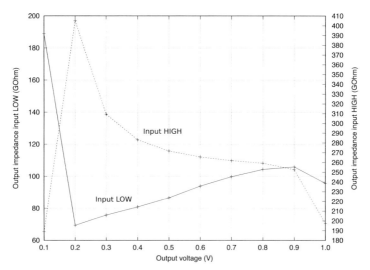

Figure 7.21: Output impedance in *highZ* state (low power tristate buffer)

124

In contrast to the reference tristate buffer, the newly implemented tristate buffer shows a different behavior. First of all, the output impedance is decreased in direct comparison to the predecessor design and in addition to this, the output curve strongly depends on the voltage at the output node and reveals a proportional dependency. The higher the voltage at *Out* is, the higher the output impedance will be. Despite the fact that the low power tristate buffer's active *highZ* curve has a smaller order of magnitude, the evaluated results are still acceptable and give an evidence about the appropriateness for the usage as a connecting element in complex designs. These results could be improved by modifying the gate lengths of the output transistors *M2* and *M3*. The downside of this modification would lead to necessary modifications of the manufacturing process, but which can be easily handled by modern technology nodes. By picking up this thought, the question for an alternative design may arise. The internal buffer of the low power tristate buffer, consisting of transistors $M2, M3, M8$ and $M9$, provides a direct signal propagation path from *In* to Out, regardless of the built-in decoupling measures from the supply voltage and *GND* and, in the worst case, vice versa. Like mentioned before, tweaking the gate length of these transistors, especially of $M8$ and $M9$ would be a suitable way to raise resistance of each MOSFET, but this should be rather treated like the last option. Hence, anticipating any external throughput on internal nodes can be achieved by swapping the transistors of the internal (and simple) core logic with the cut-off switches, shown in Figure 7.22.

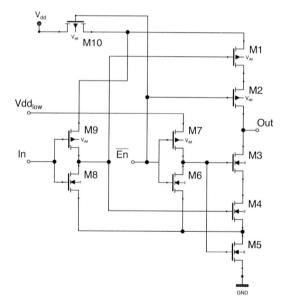

Figure 7.22: Low-power tristate buffer with swapped output transistors

Now the circuit is still similar to its predecessor shown in Figure 7.13 but features some mandatory adaptions. First of all, the internal buffer which was placed right in the centre of the predecessor's circuit before, is now split up into the transistors $M1, M4, M8$ and $M9$. So the idea here is that the second inverter of the internal buffer, implemented of $M1$ and $M4$, is now 'pulled apart' and flanked by transistors ($M2, M3, M5$ and $M10$) which serve as cut-off switches to V_{dd} and GND. All previously described energy saving measures have been applied here in the same way and simulated respectively to all tests done before. The simulation results can be seen in Figure 7.23.

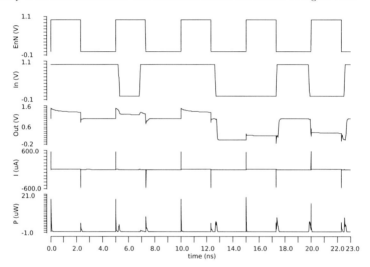

Figure 7.23: Transient analysis of modified low power tristate buffer

As expected, the modified low power tristate buffer works fine and transmits correctly all input data in normal operation mode. Activation of the *highZ* mode works fine as well, by driving the output voltage to a not specified, floating voltage value. The 'kink' of *Out* during the falling edge if *In* goes back to the fact that the input signal switches *after* right after normale operation mode was activated by $\overline{En}$ switched from $HIGH \rightarrow LOW$. Transient analysis in standby mode is pictured in Figure 7.24. An extended simulation with a transient analysis of $1ms$ is shown in Figure C.2. After the verification of the correct function, the next step was to evaluate the circuit's power loss and to compare it with the previous design.

Compared to the results displayed in Table 7.5 and Table 7.6, it can be seen that the average power consumption is around 28.6% higher than before. But it should be taken into consideration that the internal inverter ($M6$ and $M7$) is still powered by V_{dd} and not by a lower, internal supply voltage. So a few adjustments needs to be done here to figure out how this can be improved towards a lower power loss. Investigations of this circuits behavior have been made by scaling $V_{dd_{low}}$ from

126

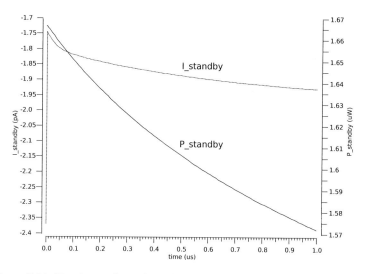

Figure 7.24: Transient analysis of modified low power tristate buffer in standby mode

the original $1V$ down to $500mV$ and are shown in Figure 7.26. This picture is a snapshot from the full simulation and all curve progressions in this figure depict a $HIGH \rightarrow LOW$ transition while leaving $highZ$ mode and entering normal operation mode. It can be clearly seen how a lower internal supply voltage of the embedded inverter impacts signal propagation. Whilst lowering the supply voltage from $1mV$ to $800mV$, the penalty in signal delay is approximately $77ps$, further decreasing of $V_{dd_{low}}$ down to $600mV$ leads to a higher penalty of approximately $622ps$. Going further down to 50% ($500mV$) of the original supply voltage, the correct function is not provided any more as a reliable inversion of In gets disrupted.

Design type	∅ P (nW)	max$\{P\}$ (μW)	min$\{P\}$ (pW)
Mod. LP tristate	268	19.81	17.69

Table 7.7: Simulation results of dissipated power P

Design type	∅ I (nA)	max$\{I\}$ (μA)	min$\{I\}$ (nA)
Mod. LP tristate	133.2	563.8	585.8

Table 7.8: Simulation results of power supply current I

However, what is encountered at this point is a typical trade-off between power consumption and operating speed. Thus, the important aspect here is to make a distinction between thinkable target application of the device during development and then to put an appropriate nucleus on

127

one these applications. Since energy efficiency comes first in the scope of this research work, the decision made was to set $V_{dd_{low}}$ to $600mV$ and to continue further investigation about the circuit's characteristics. So the next question to look up for was to see how the output impedance in $highZ$ will look like. Similar to earlier investigations, simulation were carried out to check the modified low power tristate buffers capabilities in terms of decoupling itself from a bus. The outcome of these simulations are shown in Figure 7.25. In direct comparison to Figure 7.21 the output curves of the modified low power tristate buffer show an obviously more synchronized tracing, regardless of the applied input data. Since this is a feature of an improved decoupling mechanism from the output node(s), it can be considered as beneficial characteristic. For the next step, it was interesting to see how the circuits behave and compete in worst case scenarios as described in Section 7.2. This is summarized in Table 7.4.

Type \In/Out	0V \0V	0V\1V		Type \In/Out	1V \0V	1V \1V
Ref	6.85	12.38		Ref	4.8	54.3
LP	189	95.69		LP	189	197.4
Mod. LP	211.9	94.16		Mod. LP	513.4	200.78

Table 7.9: Output impedance ($G\Omega$)

All numbers shown in Table 7.4 show an almost continuous improvement over all measurement which have been done. So by reasoning this outcome, a modified low power tristate buffer fulfills exactly the task by far, for which it was intentionally designed without raising the number of needed transistors. Introducing the architectural change in terms of placing the decoupling transistors leads to the desired effect. Additional results, showing the circuit's behavior by applying LOW and $HIGH$ as input are depicted in Figure C.5 and Figure C.6. Despite of focussing on low power consumption and high output impedance, each circuit has to transmit analog or digital data and therefore operating frequency should be also evaluated to complete the whole picture when investigating a design. For that reason, a comparison of all analyzed designs will be shown right after introducing a further variant of the tristate buffer on the following pages.

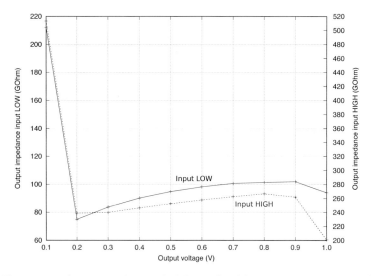

Figure 7.25: Output impedance in highZ state (modified low power tristate buffer)

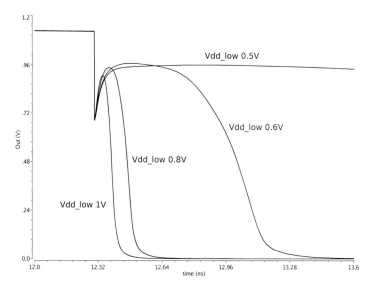

Figure 7.26: Transient analysis of modified low power tristate buffer with scaled $V_{dd_{low}}$

After completing the low-power and *highZ* evaluation of the introduced circuits, the question came up whether it is still possible to achieve further improvements with a complete redesign of the common architecture. The circuit of this new buffer is illustrated in Figure 7.27.

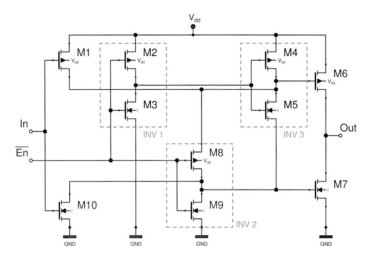

Figure 7.27: Alternative Tristate Buffer

The basic idea of this new, alternative (AT) tristate buffer is to have a separate decoupling of each of the four transistors ($M1$, $M6$, $M7$ and $M10$) which form the core buffer by additional inverters ($INV1$, $INV2$ and $INV3$). Buffer functionality is ensured by two inverters, consisting of $M1$ and $M10$ (input inverter) as well as $M6$ and $M7$ (output inverter). For normal operation mode $\overline{En}$ is turned to LOW, which leads to turning on the pMOS transistors $M2$ of $INV1$ and also $M8$ of $INV2$. So $M2$ pulls the output of $INV1$ to V_{dd} and therefore turns on $M5$ of the next inverter ($INV3$). In parallel to this, $\overline{En}$ also turns on $M8$ of $INV2$ which short-circuits the drain node of $M1$ to the drain of $M10$. The short-circuit loop which ensures this functionality is displayed in Figure 7.28.

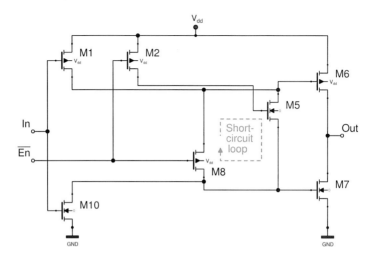

Figure 7.28: Short-circuit loop in normal operation mode

By closing the internal short-circuit loop the output node of the input inverter is connected to input node of the output inverter, which leads to the desired buffer function in case that no *highZ* mode is demanded. If $\overline{En}$ is turned to *HIGH*, transistor *M3* of *INV1* is activated and subsequently pulls the input node of *INV3* to *GND*. As consequence *M4* is turned on, pulling the output of *INV3* to V_{dd} which then turns off *M6*. At the same time, $\overline{En}$ also turns on *M9* of *INV2*, which leads to turning off *M7*. So the outcome of this procedure is that each of the output inverters' transistors is switched off separately and *Out* successfully decoupled from the tristate buffer. This is also illustrated in Figure 7.29 where all related, active transistors are displayed and all turned off transistors are faded out. These measures shall ensure a better, well balanced output impedance regardless of the input signal, which might be applied even when no data shall be transferred to the output. In addition to that, low power consumption and a fast operating frequency are also points of interest, although it is clear that a trade-off between these factors is not avoidable.

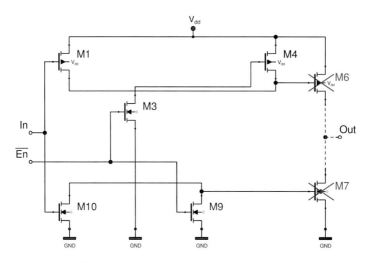

Figure 7.29: Decoupling of the output inverter in *highZ* mode

In the next step the alternative tristate buffer was simulated under the same conditions like the previous circuits and the simulation curves are shown in Figure 7.30.

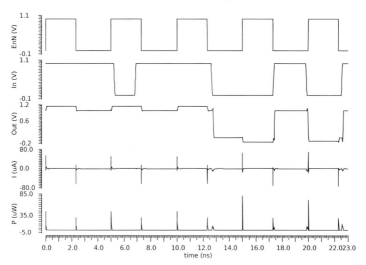

Figure 7.30: Transient analysis of alternative tristate buffer

Figure 7.30 proves the correct function as the circuit reacts correctly on the applied input data. Similar to the previous investigations, the average, maximum and minimum power dissipation was extracted from the output curves and is summarized in Table 7.10 and Table 7.11.

Design type	$\varnothing\ P$ (nW)	max$\{P\}$ (μW)	min$\{P\}$ (pW)
AT tristate	449.1	80.69	26.74

Table 7.10: Simulation results of dissipated power P of the AT tristate buffer

Design type	$\varnothing\ I$ (nA)	max$\{I\}$ (μA)	min$\{I\}$ (nA)
AT tristate	174.1	72.23	70.79

Table 7.11: Simulation results of power supply current I of the AT tristate buffer

In comparison to the results of the modified LP tristate buffer shown in Table 7.7 and Table 7.8, the newly implemented AT tristate buffer shows a poorer performance, at least when talking about a preferably low-power operation mode. The reason for this drawback is the lack of additional power reduction measures, which have been applied to the predecessors. Of course, all of these measures could have been added here as well but the basic intention was to design a completely new tristate buffer, which does not exceed a comparable number of transistors. The transient behavior of the AT tristate buffer is shown in Figure 7.31 and Figure C.7, which depicts the outputs curves with an extended duration of the simulation time. The extracted numbers are shown in Table 7.12 and underline the previous statements about the low-power capabilities of this design. A better mitigation of dissipated power and leakage current could be achieved by several add-on measures like power gating, multi-V_{th}, multi-V_{dd} and multi T_{ox}. If desired these modifications can be implemented into the AT buffer by accepting the drawbacks described earlier.

Buffer type	Mod. LP	AT
$\varnothing\ P$ (pW)	1.61	22.04
$\varnothing\ I$ (pA)	$\lvert 1.877 \rvert$	$\lvert 11.7 \rvert$

Table 7.12: Standby results comparison

After finishing the evaluation of the AT tristate buffer's energy consumption, the focus was put on investigating the *highZ* behavior. For this purpose, the circuit was simulated with an active and inactive input in order to check how this might impact the output impedance. The related output curves are shown in Figure 7.32.

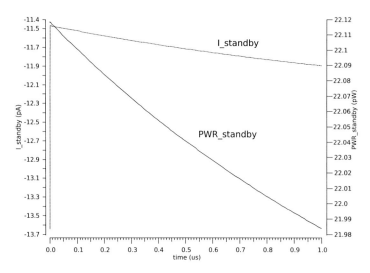

Figure 7.31: Transient analysis of the AT tristate buffer

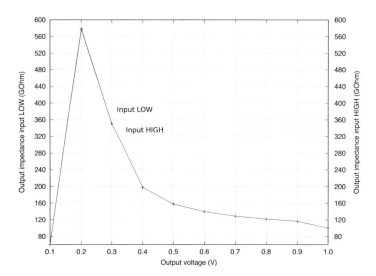

Figure 7.32: Output impedance in highZ state (AT tristate buffer)

On the first glance Figure 7.32 shows that the output curves almost match perfectly on each other. This is a desirable effect as the same output impedance is provided by the circuit regardless of the applied input. Strong, separate driving forces to cut-off each of the output transistors $M6$ and $M7$

are shown in Figure 7.29. This effect could be even reinforced by using deep cut-off measures when applying a negative V_{gs} voltage at the gates. However, the out-of-the-box result is still considerable and a comparison to the previous designs is depicted in Table 7.13.

Type $\backslash In/Out$	$0V \backslash 0V$	$0V \backslash 1V$	$\bar{x}$	*Type* $\backslash In/Out$	$1V \backslash 0V$	$1V \backslash 1V$	$\bar{x}$
Ref	6.85	12.38	9.615	Ref	4.8	54.3	29.55
LP	189	95.69	142.34	LP	189	197.4	193.2
Mod. LP	211.9	94.16	153.03	Mod. LP	513.4	200.78	357.1
AT	60.62	100	80.31	AT	60.6	100	80.3

Table 7.13: Output impedance comparison $(G\Omega)$

Similar to the previous designs, the output impedance in *highZ* mode depends on the input (*HIGH* or *LOW*) and the output voltage which will be applied from the bus to the output node of the tristate buffer during runtime. As the output impedance usually has no nonambiguous result due to its heavy dependence on the available output voltage, an arithmetic mean $\bar{x}$ was calculated. Based on the numbers shown in Table 7.13, it can be seen that the AT tristate buffer shows a decent performs when highlighting the *highZ* mode. Although its third-best arithmetic mean $\bar{x}$ it is still higher than the $\bar{x}$ of the reference design. In addition to that, the almost identical output impedance curve is one of the biggest benefits here as both thinkable cases are well equalized.

After considering all investigated circuits' capabilities for low-power operation and *highZ* performance, there is still one aspect to be further explored: the maximum circuit speed. Despite the major focus on preferably low energy consumption and the ability to provide a high output impedance if necessary, a tristate buffer should also be able to process incoming data to the external bus as fast as possible. Thus, it should be analyzed how fast a tristate buffer is able to activate and to deactivate the *highZ* mode when it's in operation. In total, four different situations should be taken into consideration at this point:

- $\overline{En} \mathrel{\widehat{=}}$ falling edge
    - *In* $1 \Rightarrow 1$
    - *In* $0 \Rightarrow 0$
- $\overline{En} \mathrel{\widehat{=}}$ rising edge
    - *In* $1 \Rightarrow 1$
    - *In* $0 \Rightarrow 0$

Rising or falling edges of the *highZ* enable signal are forbidden, therefore they are not further considered here. However, they come into account once the tristate buffer is moved into normal buffer mode. So these results provide a benchmark of the circuit's behavior in one of two possible operation modes. Thinking about the alternative, normal operation mode the typical aspects like the elapsed time for a $HIGH \rightarrow LOW$ and $LOW \rightarrow HIGH$ transition must be evaluated. All of these results are summarized in Table 7.14.

Measurement	Ref	LP	Mod. LP	AT
$\overline{En} \nearrow 1 \Rightarrow 1\ (ps)$	8.51	43.68	210	132.41
$\overline{En} \nearrow 0 \Rightarrow 0\ (ps)$	50.4	199	199	134.42
$\overline{En} \searrow 1 \Rightarrow 1\ (ps)$	7.57	2.6	919	83.94
$\overline{En} \searrow 0 \Rightarrow 0\ (ps)$	14.24	70.37	73	48.58
$t_{HL}\ (ps)$	18.47	46.7	74.78	28.48
$t_{LH}\ (ps)$	35.84	75.46	99.57	28.05
$f_{max}\ (GHz)$	18.4	8.18	5.73	17.68

Table 7.14: Timing considerations

On the first glance, Table 7.14 shows that the reference buffer outperforms each of the other designs when it comes to the maximum operating frequency f_{max}. The numbers for enabling and disabling the $highZ$ mode of the reference buffer are also considerable, which goes back to the straightforward design of it. Neither special high V_{th} transistors or stacking measures nor special high T_{ox} can be found here and therefore also do not slow down this tristate buffer during runtime. Regardless of that, the AT buffer offers a decent maximum operating frequency of $17.68GHz$ despite its modification with special measures at circuit level. This is a result which is very close to the reference buffer and goes back to just two transistors in the output inverter, shown in Figure 7.33. The maximum operating frequency f_{max} goes back to the slew rates of the rising and falling edge during each transition, which are additionally displayed in Figure C.8. One disadvantage of the AT tristate buffer is the delay in entering and leaving the $highZ$ mode, which is also listed in Table 7.14 ($\overline{En} \nearrow$ and $\overline{En} \searrow$). This goes back to the three Inverters ($INV1, INV2$ and $INV3$), which add a certain delay time t_p to active or deactivate this mode. A possibility to mitigate this negative effect would be to carefully modify the transistor sizes by accepting a possibly higher power consumption, more area consumption and more effort for the later layout of the design. Therefore these measures were not considered in this work here. For the sake of a better overview about the numbers provided in the table above, the results are displayed in Figure C.9 and Figure C.10.

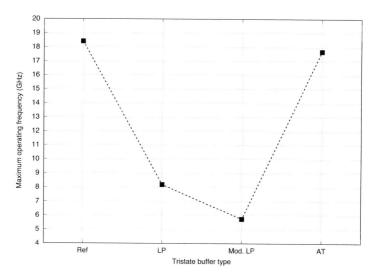

Figure 7.33: Maximum operating frequencies *fmax* of different tristate buffers

More transistors would improve the leakage current suppression, but also slows down the design while operating. Separate inverters in signal path to both transistors in the output stage add strong driving capabilities and therefore also contribute to faster switching transitions. The downside of the AT tristate buffer design can be seen by evaluating the required time to switch into the *highZ* state and vice versa. At this point, the extra logic required for the realization of the short-circuit loop takes an additional amount of time.

7.5 Simulation results

For a better comparison of the investigations which have been done, all results were summarized in Table 7.15. The low power tristate buffer outperforms in almost each aspect the reference design, which highlights its appropriateness for use in applications with limited energy resources. Results of dynamic behavior show that power dissipation is reduced significantly, no matter whether the average, maximum or minimum power consumption is in focus of discussion. The most remarkable reduction is allocated to static behavior of both circuits. Here, the standby leakage current and the dissipated power in idle mode are lowered by over 80%.

The appropriate choice of process technology due to the multi-oxide requirements as well as careful layout of transistor parameters requires special attention and allows additional improvements. However, the low power tristate buffer delivers remarkable out of the box performance without further detailed optimization. These adaptions are achieved with a small penalty in terms of

Measurement	Ref. tristate	LP tristate	Difference (%)
$\varnothing\ P\ (nW)$	245	191.3	22 ↓
$\max\{P\}\ (\mu W)$	56.75	29.72	47.63 ↓
$\min\{P\}\ (pW)$	103.8	22.36	78.46 ↓
$\varnothing\ I\ (nA)$	215.2	225.8	5 ↑
$\max\{I\}\ (\mu A)$	230.5	194.8	15.49 ↓
$\varnothing\ I_{leak}\ (pA)$	133.6	24.1	82 ↓
$\varnothing\ P_{standby}\ (pW)$	132.1	22.04	83.32 ↓
$highZ$ max.	12.38 $G\Omega$	81 $G\Omega$	↑↑
$highZ$ min.	6.85 $G\Omega$	18 $G\Omega$	↑↑

Table 7.15: Simulation results

transistor count and area. Xilinx provides 372 *Maximum User I/O* and 165 *Maximum Differential I/O Pairs* [105], which could be realized in 537 GPIOs. Implementing a new FPGA design by usage of the low power tristate buffer requires 1074 additional transistors. Here it comes to the point where an efficient layout of the overall chip could be a measures to catch up this drawback. This could be improved by a further optimization of the transistor parameters in terms of length and width. However, this might lead to a higher energy consumption and should be carefully decided case by case, depending on which characteristic is of higher importance for the respective application. Despite the additional parasitic capacitances which come along by adding transistors, nearly all measured insights does not weaken the positive overall print.

7.6 Conclusion

Each IC is of no use if it can not communicate with the outside world. The amount of GPIOs usually rises with the complexity of a chip, as big amounts of data need to be shifted into the internal registers for further processing. An essential part of GPIOs are tristate buffers, which offer the important *highZ* state to protect the stored data of internal register against any kind of overwriting by external chips using the same bus. Investigating the *highZ* mode means to find the highest impedance of a tristate buffer working in this special mode, but it should be made clear that a perfect and sole result in terms of input impedance does not exist. Instead, the impedance varies in dependance of the applied voltage at the output node. This could be put into more constant gradient by modifying the transistor parameters which are directly connected to the output node, but this could decrease the maximal operating frequency and is not a desirable consequence. Aligned to that, further research work has given evidence for the presumption that it is very difficult to find the ultimate solution, which satisfies every aspect of a design. The LP tristate buffer cuts down the power dissipation by a considerable percentage number and is even further excelled by the modified LP tristate buffer. The AT tristate buffer consumes more battery power during runtime, but delivers a significantly higher maximum operating frequency and an almost identical progression of the output impedance, regardless whether the inputs are active or inactive. An additional benefit is all achievements elaborated in previous chapters, as GPIOs contain CRAMs as well as D-FFs. These power-optimized components also contribute to a successful extension of battery lifetime.

Chapter 8

Partial component integration

This research work started with a basic analysis of a low-budget FPGA architecture. In general it can be also stated that elements of low-end FPGAs can be also found in their high-end counterparts like *Xilinx Zync*, but the downside of these products is usually the allocated costs. After redesigning and optimizing dedicated parts of a typical FPGA architecture, Figure 8.1 is used for a short recap and highlighting of the work done until now. Similar to all previous sections, all newly implemented blocks are highlighted in orange color. All PSMs are not highlighted as they have not been in focus of this work and will be addressed in future steps instead.

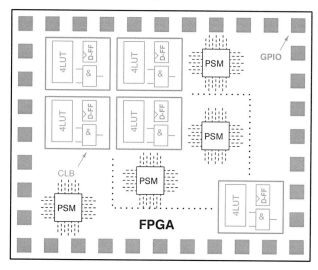

Figure 8.1: Optimized blocks of a typical FPGA architecture

So in the previous sections, selected circuits of a FPGA's internal structure were analyzed, reworked and continuously improved to mitigate the power dissipation of each considered unit. These units

141

were different SRAM cells, various D-FFs designs and a number of selected tristate buffers. All circuits have been standalone tested and their correct function verified. Different measurements have been applied to check the achievements in terms of power dissipation reduction and leakage current suppression for both, static and dynamic behavior and several more, type-specific benchmarks. Despite the fact that it was not possible to implement a whole FPGA due to lack of PSMs, it was still interesting and necessary to analyze these circuits when integrated to a component at a higher hierarchy level of a FPGA. So the next steps led to the following activities:

- Integration of SRAM designs as CRAM cells in a LUT

- Designing a SLICE by adding logic gates and D-FFs

- Implementing a GPIO by usage of the previously developed tristate buffer

These steps and the related results will be described and discussed in the following sections.

8.1 LUT design

Before stepping into the details of LUT design, it was interesting to recall the architectural elements of a *XILINX Spartan 3A SLICE*, as shown in Figure D.1. According to Figure D.1, a SLICE can be split up in two almost identical parts by mapping a horizontal borderline right in the middle of the figure. The major blocks inside a SLICE are both LUTs and the D-FFs, which are connected to each other by a set of multiplexers and logic circuitries like AND and Exclusive OR (XOR) gates. So in general it can be stated that it's sufficient to implement just one half of the symmetric SLICE block and then duplicate and appropriately wire the design for getting a full SLICE implementation. As the design of the major blocks was already handled in the previous sections, the remaining parts of the SLICE were investigated in this section in order to check whether additional achievements in terms of power savings were feasible. XOR gates are used in Figure D.1 to transmit logic signals and their functionality is described by Equation 8.1. The related truth table of this gate is shown in Table 8.1.

$$f_{XOR} = A \oplus B \tag{8.1}$$
$$= \overline{A} \cdot B + A \cdot \overline{B} \tag{8.2}$$

So according Table 8.1 the output of a XOR gate is only 1 or a logic *HIGH* if the inputs are not equal. An example implementation of a XOR gate can be found in Figure 8.2. The standard XOR design consists of eight transistors, which are equally distributed on the PUN and PDN. Depending on the applied input values, the output node will be either charged to V_{dd} or discharged to *GND*. Figure 8.3 displays the simulation results of the standard XOR design. All used transistors are standard V_{th} derivates for keeping potential transition delays as low as possible.

142

A	B	A ⊕ B
0	0	0
0	1	1
1	0	1
1	1	0

Table 8.1: XOR truth table

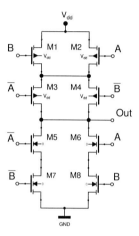

Figure 8.2: XOR gate

Figure 8.3 proves the correct function of the design and Table 8.2 summarizes the extracted results for this particular XOR design.

Design type	∅ P (nW)	max$\{P\}$ (μW)	min$\{P\}$ (pW)	∅ I (nA)	max$\{I\}$ (μA)
XOR	102.9	7.42	38.39	|99.89|	|5.73|

Table 8.2: XOR gate simulation results of dissipated power P power supply current I

For the sake of simplicity, this design was not modified with special power reduction measures, therefore it was interesting to see the circuits behavior in case that it is not driven by active input signals. This situation may come up when all inputs are set to *GND* by powering down a block which drives the XOR gate. Figure 8.4 displays the standby current and the related standby power dissipation in such a scenario during runtime. It can be seen that this circuit provides a clean transient response and engages the standby current at approximately \|31\|pA, which is acceptable for an implementation without any special adaptions for low-power applications.

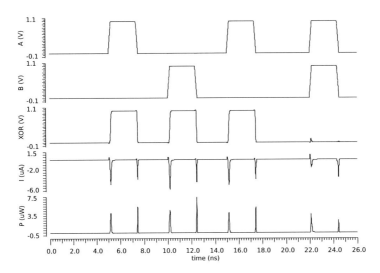

Figure 8.3: Simulation results of power dissipation of the standard XOR design

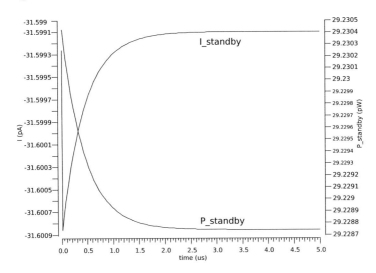

Figure 8.4: Transient analysis of the standard XOR gate in standby mode

Furthermore, the internal stacking effect of the stacked transistors due to the design principle of this XOR gate, supports leakage current flow. Despite this design feature, the most significant drawback here is the chance of a current flow from V_{dd} to GND. This might happen during switching events while on normal operation and in standby mode, leading to undesired leakage currents which drain battery power over time. Thus, the overall results for both, dynamic and static power consumption, still could be improved by choosing an alternative design approach. A possible solution is to build up a XOR gate by using a 2:1 TG MUX [120]. So the basic idea here is first of all the decrease to the number of transistors and to limit the number of transistors in the leakage current path as well. This implementation is shown in Figure 8.5.

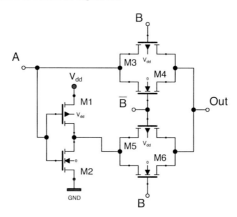

Figure 8.5: XOR gate with 2:1 TG MUX

As shown in the figure above, the XOR function is realized by a 2:1 MUX network which is driven by the input signal A and controlled by B and $\overline{B}$. The latter signals are used to open or to close both TGs. The simulation results of the dynamic behavior for both XOR gate designs are summarized in Table 8.3 and are additionally shown in Figure 8.6.

Design type	$\varnothing\ P\ (nW)$	$\max\{P\}\ (\mu W)$	$\min\{P\}\ (pW)$	$\varnothing\ I\ (nA)$	$\max\{I\}\ (\mu A)$
XOR	102.9	7.42	38.39	|99.89|	|5.73|
MUX XOR	99.84	6.694	93.73	|89.58|	|7.419|

Table 8.3: Simulation results of MUX XOR gate (P and I)

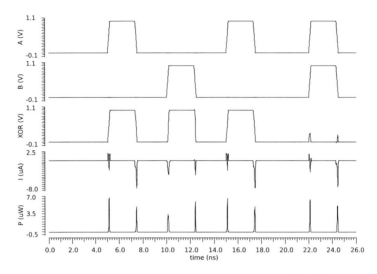

Figure 8.6: Simulation results of power dissipation of the MUX XOR design

Minor improvements can be seen in Table 8.3, but the most significant advantage at this point is certainly savings in area consumption as only six transistors are required to implement the circuit. In addition to that, the standby behavior was also investigated in a simulation by switching off all inputs. The simulation curves for that particular case are shown in Figure 8.7. This figure shows at a first glance that both, leakage current and related leakage power dissipation, are significantly smaller than before. The transient responses of both measurements are finished in a significantly less amount of time than before, which goes back to a smaller number of transistors in the signal path and therefore less delay as well. So in this case, the standby current $I_{standby}$ engages at approximately $|6.4|\mu A$ and is therefore remarkably lower than the standby current of the standard XOR gate.

One last question to be answered during the research on XOR gates was whether the dynamic power dissipation as well as the transistor count could be further decreased. The answer on this question is displayed in Figure 8.8. This shown TG XOR gate consists of a TG gate and a pair of MOSFET transistors only, leading to a decreased transistor count of four. Furthermore, it shall be stated here that the output node *Out* is not supplied by V_{dd}, therefore the driver capabilities of this design might be limited in comparison to the latter implementations. As a consequence, the input signals B and $\overline{B}$ must supply all necessary current to drive the output capacitance. On the other hand, avoiding to connect this gate to V_{dd} and to GND prevents leakage currents from the power supply to GND as this possibility is not given any more. A precondition for the correct function of the TG XOR is full swing voltage support of B and $\overline{B}$, otherwise correct signal propagation to

146

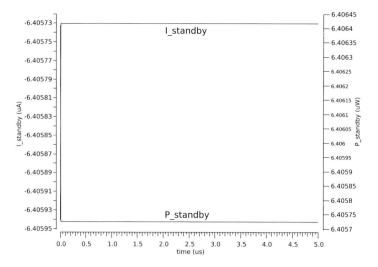

Figure 8.7: Transient analysis of the MUX XOR gate in standby mode

following logic blocks might not be guaranteed due to limited driving force. The simulation results are shown in Table 8.4 and also displayed in Figure 8.9 and Figure 8.10. The results in Table 8.4 show clearly that the TG XOR gate outperforms the previous XOR derivations in each evaluated aspect. The average power dissipation is cut down by approximately 85%, which is a considerable result. One additional evaluation to be done was a comparison of achievable power savings and leakage current suppression in standby mode.

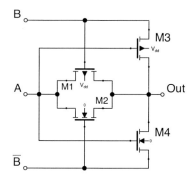

Figure 8.8: TG XOR gate

The respective results of this evaluation of all analyzed circuits in a steady state are listed down in Table 8.5. In addition to that, f_{max} was also evaluated and added to this table, as the maximum

147

Design type	∅ P (nW)	max{P} (µW)	min{P} (pW)	∅ I (nA)	max{I} (µA)
XOR	102.9	7.42	38.39	|99.89|	|5.73|
MUX XOR	99.84	6.694	93.73	|89.58|	|7.419|
TG XOR	14.64	2.868	18.19	|0.057|	|3.138|

Table 8.4: Comparison of all simulation results of P and I

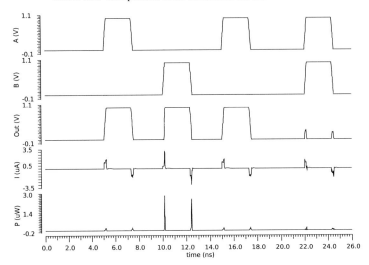

Figure 8.9: Simulation results of power dissipation of the TG XOR design

achievable operating frequency should not be neglected and should still be considered. The explicit results shown there underline the fact that TG XOR gates can be seen as reasonable choice for the implementation of low-power architectures in reconfigurable logic.

Measurement	XOR	MUX XOR	TG XOR
∅ $P_{standby}$	29.23pW	6.406µW	3.046fW
∅ $I_{standby}$	|31.6|pA	|6.406|µA	|3.99|pA
f_{max}	6.8GHz	3.1GHz	3.12GHz

Table 8.5: Simulation results of $P_{standby}$, $I_{standby}$ and f_{max}

So in general it can be stated that the TG XOR provides decent results in terms of power dissipation and therefore qualifies itself to be used in applications with limited energy resources. Of course, there are some drawbacks which have to be considered when designing a chip for special purposes. Standard XOR gate implementations offer a higher operating frequency and better driver capabilities, but can not compete with TG XOR gates when it comes to power considerations and area

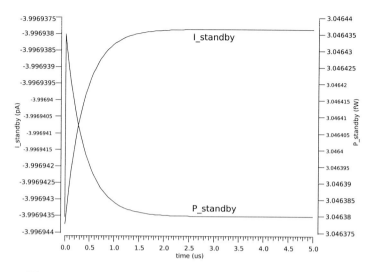

Figure 8.10: Transient analysis of the TG XOR gate in standby mode

consumption. Further improvements could be achieved by using high V_{th} *MOSFETS* to design a TG XOR gate. All previously described power saving measures could be also applied to this logic gate in order to get even better results.

Beside the XOR logic a SLICE also contains AND gates. This kind of logic is very simple, but should be also considered when implementing a whole SLICE. Two selected designs for an AND gate are shown in Figure 8.11. The AND gate on the left hand side works well, but it also reveals a significant weakness in its internal design: pMOS transistors are used for the PDN and nMOS transistors are used for the PUN. Especially in case that high V_{th} transistors are used in the PDN, a $V_{gs} = 0V$ will be not sufficient to operate the pMOS transistors in saturation region, but in subthreshold region instead. This leads to a reduction of usual current flow and therefore in slower switching times during operation. Decreased current flow can be a desirable effect when it comes to focus on reducing energy loss due to unintended power dissipation in static mode. If the same effect appears in dynamic behavior, then the consequences might be unfortunate as the whole circuit slows down its speed significantly. A solution to overcome this problem would be to use negative V_{th} voltages so that $V_{gs} = 0V$ will be sufficient to fully turn on the transistor. Alternatively, the circuit on the right-hand side of Figure 8.11 can be chosen to overcome the previously described problems. This design is basically a standard 2-input NAND gate which is connected in series to a following inverter. Despite the fact that this solution requires two additional transistors, the main advantage here is that pMOS transistors are used to realize the PUN and nMOS transistors are used to realize the PDN. This circuit can be extended by using the stacking effect in the PDN of the embedded NAND gate, which could be taken into consideration when working on further power saving measures. The respective simulation results, including a comparison of the output curves and supply current are displayed in Figure 8.12.

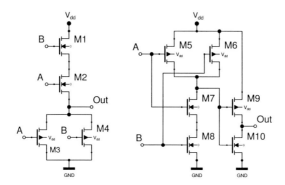

Figure 8.11: Simplified AND gate (left) and improved AND gate (right)

According to Figure 8.12, it is obvious that the simplified AND gate does not achieve full swing output over runtime. Signal $AND\_simplified$ shows that an unambiguous LOW level can not be achieved, for example during a $00 \Rightarrow 10$ transition. Taking a closer look on the output of a $00 \Rightarrow 11$ transition also reveals that the simplified AND gate does not manage to achieve full power supply level of $1V$, as it ends up with a maximum voltage level of approximately $700mV$. In contrast to this, the improved AND gate offers a way better operating behavior, which is proven by taking a look on the output curve $AND\_improved$. Similarly to the XOR gate, these gates here have also been developed by using standard V_{th} transistors only in order to avoid as much signal propagation delay as possible. The standby behavior of the circuit is shown in Figure 8.13.

Design type	$\varnothing$ P (nW)	max$\{P\}$ (μW)	min$\{P\}$ (pW)	$\varnothing$ I (nA)	max$\{I\}$ (μA)
AND	54.85	14.86	33.81	|51.98|	|16.98|

Table 8.6: Comparison of simulation results of P and I for the improved AND gate

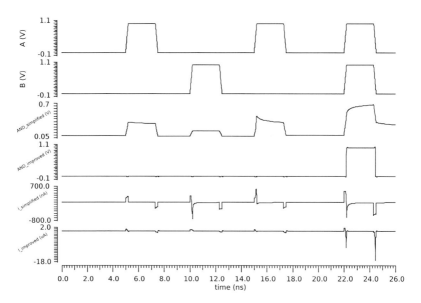

Figure 8.12: Simulation results of both AND gate designs

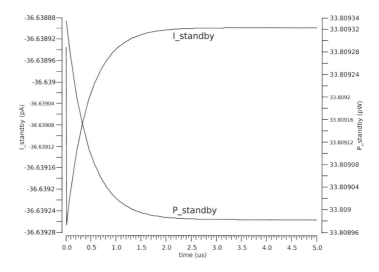

Figure 8.13: Transient analysis of the improved AND gate in standby mode

Referring back to Figure A.2, a 4-input LUT including a write and output buffer is depicted in

151

Figure 8.14. A LUT with 4 input signals, A, B, C and D, leads to the necessity of 16 CRAM cells inside it as 4^2 different combinations of input values might come up. A multiplexer tree connects the CRAM cells to the output buffer and is driven by the input values and their complements. So the MUX tree is a straightforward design element and its basic function is depecited in Figure 8.15.

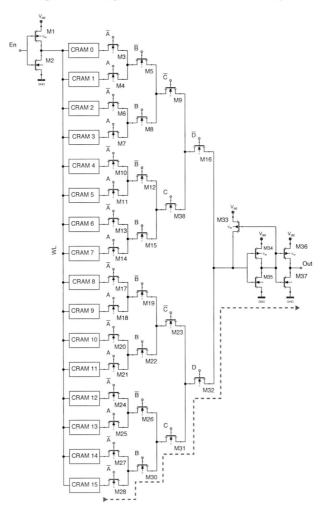

Figure 8.14: Simplified 4-input LUT with input and output buffer [48]

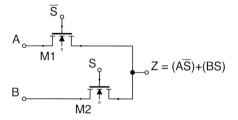

Figure 8.15: Basic function of a MUX including select inputs

In dependence of the applied *Select* signals S and $\overline{S}$ to the circuit shown in Figure 8.15, either input A or B will be transmitted to output Z. This principle is applied 1:1 in the LUT shown in Figure 8.14, where a 4-level MUX tree was implemented. For the sake of low signal propagation delay, all transistors of the MUX tree are standard V_{th} variants, except those of the first MUX tree level. These nMOS transistors are a significant of each *CRAM* cell as they form one of two access transistors of each cell. Usage of high V_{th} could be also taken into consideration, but despite the major focus for lower power applications, signal propagation should not be degraded too much at this hierarchical level. On the opposite of each *CRAM* cell, signal *EN* guarantees access via an inverter to the word line *WL* for writing configuration data into the cells. Of course, an additional write line signal WR_i must be available to each *CRAM* individually. Last but not least it should be mentioned here that all configuration cells are also connected to a five transistor cell sense amplifier, which is not shown in Figure 8.14 but in Figure A.1 instead. In order to determine the critical signal propagation path, it was important to take the 4-level MUX tree into consideration. Input signal A is the most critical signal among the others, at it drives the first level of the MUX tree and is also highlighted by the red dashed line in Figure 8.14. The worst case in this architecture might come up when a total change of all input signals appears over runtime, e.g., A, B, C and D switch from 0000 $\Rightarrow$ 1111. In addition to that it is assumed that all *CRAM* store alternating data, which means $CRAM0\hat{=}0$, $CRAM1\hat{=}1$, $CRAM2\hat{=}0$ and so on. This situation leads to the fact that each drain or source node of the transistors in the MUX tree must be either charged or discharged, which does not only lead to power dissipation but also to propagation delay. The output buffer is a slightly modified version as it consists of five transistors instead of four. While logic data propagates through the MUX tree, its voltage level degrades slightly in case that a logic 1 is transmitted. The reason for this degradation are the nMOS transistors which will lower the voltage of transferred data to approximately $V_{dd} - V_{th_n}$ until it arrives at the output node of the MUX tree. In consequence, this circumstance might become a problem for correct signal propagation by the following output buffer as the decreased voltage applied to the input of the first inverter of the output buffer might be too late to activate $M35$ fast enough. Figure 8.16 displays the described situation and compares the output curves of a standard output buffer and an improved one.

This behavior results in even more propagation delay which is quite unfortunate at that moment.

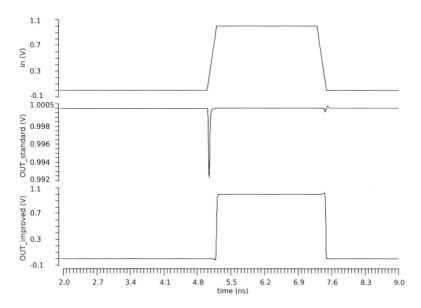

Figure 8.16: Output comparison of standard and improved buffer type

A solution for this problem is the insertion of an additional pMOS transistor $M33$ in a feedback loop line from the output of the first inverter to its input. $M33$ helps to restore the input voltage as it charges this node to V_{dd} and therefore reinforces the inverting function of $M34$ and $M35$. Similarly to the steps done in previous chapters, the straightforward strategy would be to replace all standard V_{th} transistors by their high V_{th} derivates. In this special case, turning $M35$ and $M36$ into high V_{th} transistors would further worsen the consequences of voltage level degradation caused by the MUX tree: a higher V_{th} voltage makes it even more difficult for a lowered V_{gs} to turn on a transistor appropriately. Thus, a reasonable trade-off here was to implement $M33$, $M34$ and $M37$ as high V_{th} devices and $M35$ and $M36$ as standard V_{th} derivates. The significant simulation results are shown in Table 8.7 and reflect the achievements of the improved output buffer in contrast to the standard implementation.

Design type	$\varnothing\ P\ (\mu W)$	$\max\{P\}\ (\mu W)$	$\varnothing\ P_{standby}\ (pW)$	$\varnothing\ I_{standby}\ (pA)$
Standard	224.6	13.06	76.54	|79.36|
Improved	0.89	10.77	13.37	|16.2|

Table 8.7: Simulation results of P and I for the standard and improved buffer

Especially the massive improvements in terms of leakage power reduction in standby mode contribute to decent reduction of energy waste in the circuit's idle state. After the implementation of all supplementary components, the next subsequent step was to check the low-power performance of a fully implemented LUT. For this purpose, two different LUT implementations were done: a 6T CRAM based design and the optimized counterpart, which is based on the LP 4T CRAM cell. As a LUT is the most configurable element inside a FPGA, a transient analysis with a particular configuration is not meaningful. Instead of this, a comparative benchmark is both LUTs leakage current suppression capabilities in standby mode. Switching a whole FPGA into standby mode means to put all *WL* lanes to *LOW*. The reason behind this is to decouple all CRAM cells from the MUX tree, so that no currents have the chance to leak out of the CRAM cells. Appropriate control measures for this feature need to be added to the address buffer, which was not in scope of this research work. As these cells are cut off from the bitlines, this comparison can be performed regardless of all possibly upcoming input data. Figure 8.17 shows the transient response of the 6T LUT during a 1*ns* transient simulation.

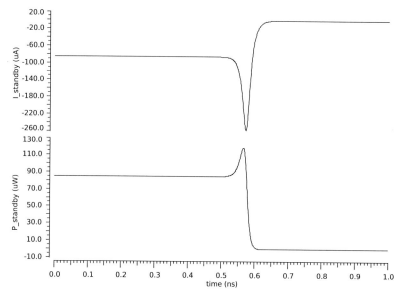

Figure 8.17: Transient analysis of the *6T LUT* in standby mode

Figure 8.17 displays the expected kink in the simulation curves, which goes back to the discharge of parasitic capacitances allocated to internal nodes of the design. Taking a look on the y-axis of both, the standby current and standby power, it can be clearly observed that both curves are within the range of μA and μW. These results were expected as the standard 6T CRAM cells do not imply any special leakage reduction measures. Moving forward to the next step the same simulation was applied on the optimized, LP 4T CRAM cell based LUT implementation and the respective results are listed in Table 8.8.

LUT type	∅ $P_{standby}$	∅ $I_{standby}$
6T	50.48 μW	$|55.79|$ μA
LP 4T	76.86 pW	$|90.44|$ pA

Table 8.8: Summarized results of standby behavior ($P_{standby}$ and $I_{standby}$)

Unambiguous numbers in Table 8.8 reveal the strong impact of energy saving LP 4T CRAM cells on a LUT design. Whilst a 6T CRAM LUT design offers a standby power of $50.48\mu W$, building in LP 4T configuration cells leads to an efficient descent of the previous result down to $76.86pW$. The shift from μW to pW leads to a massive power loss reduction of 2 orders of magnitude. It should be kept in mind that all results were determined in idle state in which the operation of a LUT is just stopped, but power gating is a different story. The LP 4T CRAM cell shown in Figure 5.30 provides a dedicated power down feature by offering the option to switch to a lower power supply V_{dd_L}, which fulfills the requirement of fine-grain supply voltage scaling. Going further to a coarse-grain level requires the ability to power down a whole LUT or even SLICE. Therefore both LUT designs were modified by an additional gating transistor in the power supply path and tested again. The respective results can be found in Figure 8.18.

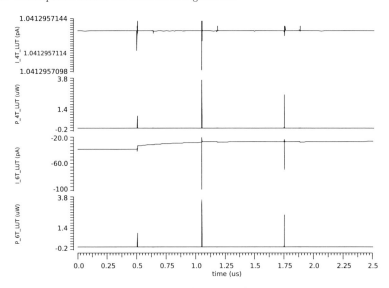

Figure 8.18: Leakage current and power dissipation of the 6T LUT and LP 4T LUT in power off mode

So Figure 8.18 shows the further diminishment of the leakage current by cutting off the power supply from the whole LUT. Whilst the 6T CRAM LUT exhibits a settlement from $|55.79|\mu A$ to

156

$|28.29|pA$, the LP 4T CRAM LUT goes even further by pushing down the leakage current from $|90.44|pA$ to $|1.041|pA$. That relation shows again the LP 4T CRAM cell in use to line out leakage current flows. This capability is paramount in comparison to the standard memory cell used for configuration. In an extreme case, even the MUX tree could be implemented by using high V_{th} transistors, but the improvements which came up by leaving the MUX in standard V_{th} configuration are considered to be sufficient for being seen as achieved design goal at this level.

8.2 SLICE design

The next step was to extend the previously developed LUT by additional circuit blocks to implement a SLICE like shown in Figure D.1. As each SLICE basically consists of two major blocks (separated by the dashed borderline in Figure D.1) connected to each other, only one of these blocks was designed in this work. This block and its modified components (highlighted in orange color) are displayed in Figure 8.19. It should be stated here that this illustration is a simplified of the real internal structure. For example, by taking a closer look on Figure D.1, special components like multiplexers without a dedicated *Select* input can be seen, shown in Figure 8.20. The modification of the partial SLICE in scope was done by a stepwise approach, starting by the replacement of the standard 6T LUT against an optimized LP 4T LUT variant. This step was followed by replacing the logic gates and the D-FF by the newly developed, energy-saving counterparts.

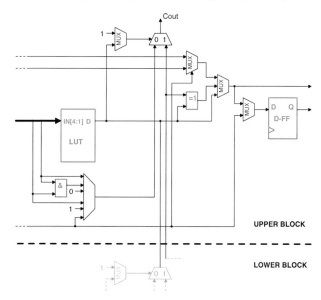

Figure 8.19: Partial SLICE structure of an *XILINX SPARTAN 3A* FPGA

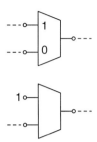

Figure 8.20: MUX components without dedicated *Select* input

These multiplexers in Figure 8.20 are usually configured by the toolchain in use, which means that the *Select* inputs are determined by the tool which streams the bitstream of the intended design into the FPGA. The upper MUX shown in that figure will forward either the upper or lower input, depending on whether *Select* will be configured to be *HIGH* or *LOW*. The lower MUX displayed in the same figure is another case: depending on the pre-configured *Select* input, either the static, pre-initialized $'1'$-input will be put through to the output of the MUX or the dynamic input. Depending on the configuration done by the toolchain, a vast number of different SLICE configurations is feasible. Based on that fact, measurements of leakage current and dissipated power during standby and power-off mode was determined and compared. The output curves of these simulations can be seen in Figure 8.21.

On the first glance, Figure 8.21 reveals a supposed delay in the transient response of the optimized SLICE while in idle state. It is not a real delay though as there are apparently different orders of magnitude between the standard and optimized design (μA vs. pA and μW vs. nW). The amplitude of the transition to be found in the output curve of $I_{standby_{LP}}$ is smaller than of $I_{standby}$, so the alleged delay (which takes just approximately $1.5ns$) is no one. Also, by having a closer look on the y-axes of the output curves, it is obvious that the scaling of the low-power outputs ($P_{standby}$ and $P_{standby_{LP}}$) is different between both simulation. The same fact also applies to the consideration of leakage current in standby mode, which can be considered as an outlook on the overall results of this simulation. So, referring to more detailed numbers, Table 8.9 delivers the required information.

$SLICE\ type$	$\varnothing\ P_{standby}$	$\varnothing\ I_{standby}$	$\varnothing\ P_{power_{off}}$	$\varnothing\ I_{power_{off}}$	N_{MOSFET}
Standard	48.65 μW	$|48.32|\ \mu A$	2.878 nW	$|29.88|\ pA$	182
Low-power	1.722 nW	$|173.3|\ pA$	934 pW	$|16.41|\ pA$	213

Table 8.9: Summarized *SLICE* results in standby and power-off mode

Table 8.9 shows a massive improvement in terms of power savings by reducing power dissipation from $48.65\mu W$ down to $1.722nW$ during standby of the SLICE, which is a quite considerable result.

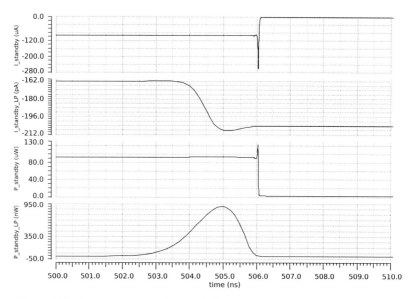

Figure 8.21: Leakage current and power dissipation of the standard SLICE and the *LP SLICE* in standby mode

All previously applied improvements show their expected impact at this hierarchical level and lead to a tremendous reduction of current drain from limited energy resources. Going further by adding power gating to the supply voltage, further mitigation of undesired leakage currents are realized. This can be seen in Figure 8.22, where the effect of cutting off V_{dd} from the circuit by setting *En* to *LOW* is demonstrated. Table 8.9 also proves that this measure helps the partial standard SLICE to show up power savings in power off mode as well, but still can not beat the achievements of the optimized SLICE when completely turned off.

These achievements come at cost of a higher transistor count, leading to a difference of 17%. Beside the necessity of putting more attention to the layout of the design due to balancing out parasitic capacitances, the higher number of transistors is the most significant disadvantage of the newly implemented partial SLICE.

159

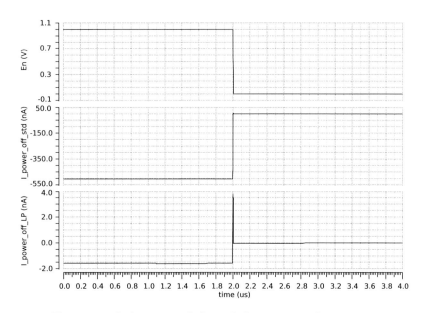

Figure 8.22: Leakage current before and after activation of power gating

8.3 GPIO design

The last, concluding step was to refer back to the introduced tristate buffer in Section 7.4. As described in that section, a number of tristate buffers was analyzed upon their capabilities to, e.g., save as much battery power as possible in static and dynamic mode. In that sense, it was also of interest to explore the power savings capabilities when embedding a tristate buffer into a GPIO. For that reason, Figure 7.3 is picked up again in this chapter and redrawn to highlight in orange color the modified parts of an low-power optimized GPIO. This is shown in Figure 8.23.

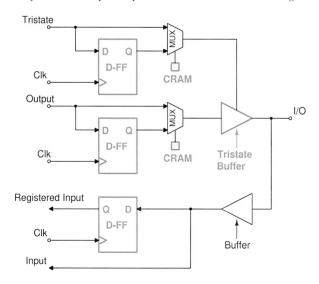

Figure 8.23: Simplified GPIO structure with highlighted, optimized blocks

At this point, it was beneficial to take advantage of the previous work steps, as a GPIO implies two CRAM cells to configure its internal multiplexers and three D-FFs, which are used to store input and output values, if necessary. Based on that fact, the consequent step was to use the low-power optimized parts from Section 5.2, 6.3 and 7.4 and to integrate them into the GPIO architecture. Figure 8.23 features three D-FFs in total, hence the CR D-FF of Section 6.3 was chosen to be applied in an energy saving GPIO. Next, two CRAM cells are required to configure the *Select* inputs of the multiplexers, so it was self-evident to implement the LP 4T CRAM cell for this purpose. Last but not least, the tristate buffer itself was realized by taking credit of one of the modified designs from Section 7.4. The modified low-power tristate buffer with swapped output transistors shown in Figure 7.22 was chosen for the sake of power consumption reduction as it provides decent results, which were matching best to what was intended here. Similar to the previous section, the most interesting part was the standby and power off leakage current flow and power consumption. For the sake of a more detailed comparison, all simulations were done with all kinds of before developed tristate buffers. So for the first run, all GPIO variants were analyzed upon their capability to suppress leakage current and the related power dissipation during standby. The standard GPIO design is used as reference implementation and its leakage current and related power dissipation can be seen in Figure 8.24, followed by a comparison with the other designs.

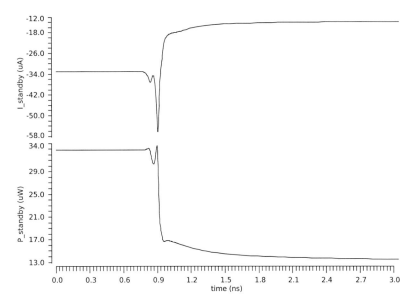

Figure 8.24: Leakage current and dissipated power of the standard GPIO in standby mode

Output curve of $I_{standby}$ in Figure 8.24 shows the expected transient response which goes back to internal capacities. One remarkable point at this is place is the order of magnitude of both, $I_{standby}$ and $P_{standby}$ which is in the order of μA and μW, even after transient response has reached a stable value (from $1.8ns$ simulation time onwards). The reason for this relatively high order of magnitude is lacking of dedicated measure against the root cause of this outcome. In consequence, it was of high interest to see how the alternative, low-power designs would cut down leakage currents. The related results of $I_{standby}$ of all four investigated designs can be seen in Figure 8.25.

162

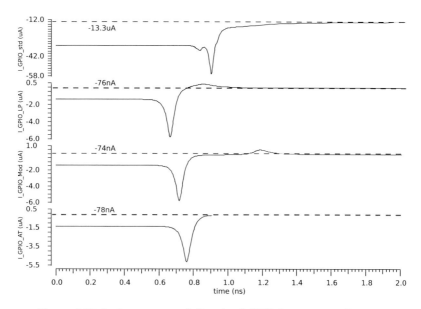

Figure 8.25: Leakage currents of all analyzed GPIO designs in standby mode

According to Figure 8.25 all alternative designs, the LP, Mod. LP and AT GPIOs provide a decreased standby current when compared to the standard design. All of the alternative, low-power designs show a good performs in terms of leakage current reduction, but the Mod. LP GPIO is best though. Going further into detail, all related results during standby mode are summarized in Table 8.10.

GPIO type	$\varnothing\ P_{standby}$	$\varnothing\ I_{standby}$	$\varnothing\ P_{power_{off}}$	$\varnothing\ I_{power_{off}}$	N_{MOSFET}
Standard	16.04 μW	$|16.32|\ \mu A$	38.30 pW	$|39.36|\ pA$	78
LP	853.8 nW	$|64.73|\ nA$	4.199 pW	$|5.203|\ pA$	87
Mod. LP	673.7 nW	$|51.64|\ nA$	4.186 pW	$|5.192|\ pA$	87
AT	1.010 μW	$|287.0|\ nA$	4.377 pW	$|5.381|\ pA$	87

Table 8.10: Summarized GPIO results in standby and power-off mode

Referring to Table 8.10, it can be seen that the modified LP GPIO offers the lowest average standby power dissipation $P_{standby}$ and average standby leakage current $I_{standby}$. The number of transistors N_{MOSFET} is equal to the other, alternative designs and only makes a difference when taking the standard GPIO into consideration. Here, the more complex implementation of the integrated D-FFs, CRAM cells and tristate buffers leads to an increased number of transistors. The AT tristate buffer based GPIO falls behind the power reduction performance of the other, newly implemented

163

designs, but this result was expected according to the elaborated findings in Table 7.12. Figure 8.26 illustrates the differences between the results in a graphical way, highlighting the effect of power loss mitigation after applying dedicated measures at circuit level. In addition to that, the average leakage current $I_{power_{off}}$ and the correlated, average power loss $P_{power_{off}}$ was measured and added to Table 8.10. The respective numbers for $I_{power_{off}}$ and $P_{power_{off}}$ in this table underline the fact that the low-power optimized design outperform the standard design. It shall be stated here that other measures, e.g., clock-gating could be also applied in case that power gating is not an option due to the necessity to keep the data stored in the CRAM cells and D-FFs.

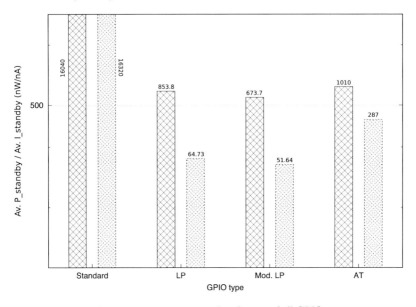

Figure 8.26: $\varnothing\ P_{standby}$ and $\varnothing\ I_{standby}$ of all GPIOs

However, clock-gating was not taken into consideration during this evaluation as it can not be applied on the reference D-FF. According to the overall decent results, the modified LP tristate should be the first choice when it comes to designing a low-leakage GPIO circuit. A final illustration for showing up the different transient responses in standby mode of the standard GPIO and the modified tristate buffer LP GPIO is displayed in Figure D.2.

8.4 Conclusion

The integration of components, which have been developed in Chapter 5, 6 and 7, shows an overall positive effect in terms of power consumption reduction. All low-power components contribute successfully to considerable power savings. Despite the fact that these results are limited to the static behavior of the circuits in scope, it can be still stated that embedding power saving measures at the lowest hierarchical level is a reasonable decision. Each integrated component, the LUT, partial SLICE and the GPIO take credit of their power-optimized subcomponents and provide smaller leakage currents. This fact has an immediate impact on the overall power dissipation and qualifies each integrated component for low-power applications. It also proves that it is reasonable to optimize even the most basic logic blocks of a design, as every contribution in low-power application counts.

Chapter 9

Summary / Outlook

9.1 Summary

Configurable logic provides interesting opportunities for many applications which would otherwise take many months of development, e.g. fast prototyping. Reconfiguration also offers the chance to load new functions and to replace those, which are not necessary any more. Battery-powered systems would benefit from this feature as reconfigurable logic may replace a number of dedicated logic. However, FPGAs like the Xilinx Spartan 3A do not comprise dedicated power-saving measures at the lowest circuit level. Related software tools support insertion of typical design techniques like pipelining, but do not have any impact on the schematics as the FPGA is already fabricated at that point of time. Taking this face as a baseline, selected parts of the Xilinx Spartan 3A were replaced by their leakage current optimized equivalents. Priority was put on static power dissipation, but dynamic power dissipation was also considered during the optimization work.

At first, an improved SRAM cell was developed which serves as CRAM cell inside a LUT, described in Section 5.2. Despite the fact that the legacy designs were also equipped with the same power saving measures, the newly developed LP 4T CRAM cell outperformed them in terms of power savings by providing a slightly better SNM and comparably good WNM. The downside of these achievements is the need for additional transistors, which make the LP 4T CRAM cell more complex than the standard 6T SRAM cell.

Next, selected D-FFs were analyzed and compared against a new design. This newly developed CR D-FF, which is introduced in Section 6.4, is based on a dynamic and differential SABL style and involves an efficient combination of static and dynamic power consumption reducing measures. This leads to second best results in terms of average power savings a still a decent maximum operating frequency. In addition to this, very good defense against DPA attacks is also featured by keeping the difference of dissipated power during each switching transition at a minimum level. This design can be used in other circuits as well and is not limited to a usage within FPGAs. These capabilities come at cost of higher transistor count and higher efforts to layout this design, as parasitic capacitances must be balanced out between both outputs.

167

The last optimized part at this hierarchy layer was the tristate buffer, which is an inevitable component of Gpios and described in Section 7.4. In total, three different versions of an optimized tristate buffer were developed, but simulation results certify that the best overall low-power performance is the modified LP tristate buffer. Similar to the previous sections, the same optimization strategy was also applied here and considerable low-power improvements were achieved. One of the most important functions of a tristate buffer is to offer a decent *highZ* performance for different voltage levels which may be applied to a tristate buffers output. Here, the modified tristate buffer outperforms the other designs, while still showing good low-power capabilities. Nevertheless, the disadvantage of the modified LP tristate buffer is a lower maximum operating frequency.

After finishing the improvements of the selected blocks, in Section 8.1 integration steps were made to see the outcome at a higher architectural level. For doing so, additional optimizations were applied to basic logic gates for keeping the dissipated power as low as possible. A SLICE of an FPGA was designed partially, as it consists of two identical blocks which are connected to each other. The results in this section show that a partial LP SLICE cuts down average, static power dissipation by a magnitude which is an acceptable result. This is valid for both, standby mode and power-off mode of the partial SLICE. As expected, this low-power implementation requires more transistors and therefore also a higher area consumption.

In Section 8.3 a GPIO was implemented. At this point, it was beneficial that a GPIO also uses CRAM cells to configure its internal multiplexers and D-FFs to store input and output data, if needed. Similar to the results after realizing the partial SLICE, the LP GPIO also shows the smallest average power dissipation. Furthermore, this result can be improved by applying power-gating to the circuit, which can be similarly realized for other designs.

In sum, usage of all low-power parts and their integration to higher level components is very beneficial for realizing a power-optimized low-budget FPGA. It should also be mentioned here that all of the benefits will come along with some disadvantages. More transistors are required and a special process technology, which provides the necessary multi-V_{th} and multi-T_{ox} options. In addition to that, these drawbacks may have a negative impact on the yield during production, as additional process steps may lead to undesirable defects within the wafer. A higher number of transistors increases chip area and therefore also the yield. On balance, LP FPGA in the low-budget area (Xilinx Spartan 3 and Spartan 6 series) will not deliver high performance in terms of computational power, so a reasonable weighing up must be done to make the right choice for the appropriate application.

9.2 Outlook

For future work, all remaining parts of an FPGA should be also designed and optimized for cutting down their power consumption. As mentioned in Section 8.1, a low-power related development of the PSMs would significantly contribute to a further decrease of both, static and dynamic power consumption (see Figure 3.4 and Figure 3.5). Furthermore, the sense amplifiers and address buffers, which are required to configure and read the configuration memory cells, should be also analyzed as possible power savings can be also expected here. All research and implementation work done until this point excluded the layout of each schematic. Each layout will add more parasitic capacitances, which may distract proper signal propagation. So a layout vs. schematic (LVS) check should be performed in order to check, whether the simulation results are still compliant to the results from the pure schematic entry. Another interesting aspect to be taken into consideration is to apply a design shrink by synthesizing all designs with a more advanced process technology. Even better power savings can be expected by this measure, but other physical effects, e.g., the short-channel effect, might be something worth considering. By having all low-power optimized FPGA components available, the next step would lead to a gradual integration and verification for correct function. This newly developed FPGA could then be tested with different open source cores and compared against the results provided by commercial FPGAs.

Appendix A

CRAM - Additional Figures and Simulation Results

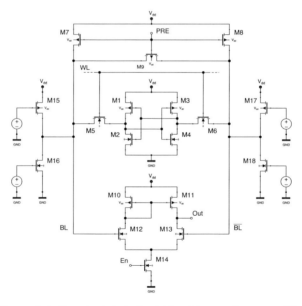

Figure A.1: Test circuit including bitline drivers and sense amplifier

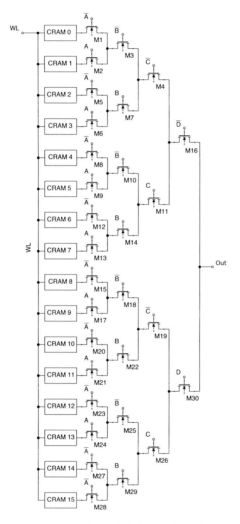

Figure A.2: Simplified 4-input LUT

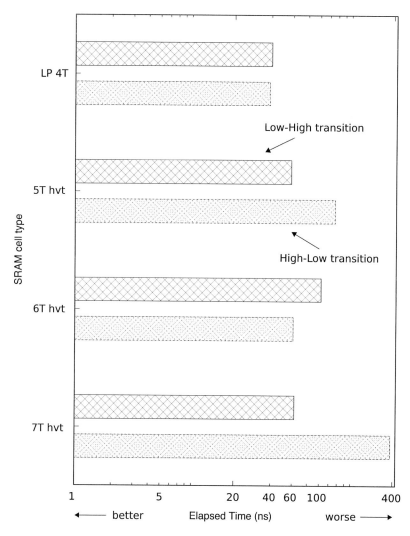

Figure A.3: Slew rates and f_{max}

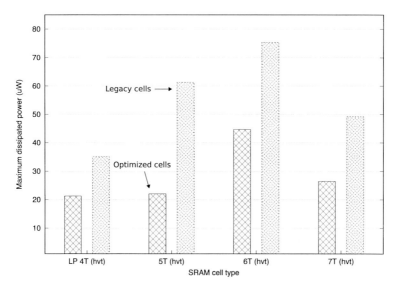

Figure A.4: Comparison of maximum power dissipation between original and modified cell designs

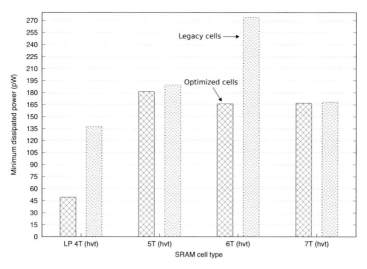

Figure A.5: Comparison of minimum power dissipation between original and modified cell designs

Appendix B

D-FF - Additional Figures and Simulation Results

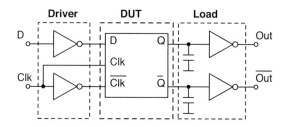

Figure B.1: D-FF test circuit

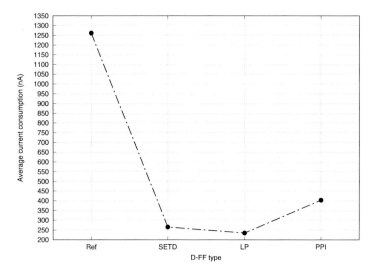

Figure B.2: Average consumed current $I_{V_{dd}}$

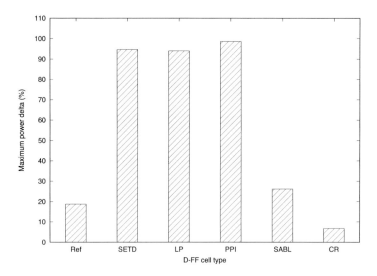

Figure B.3: Comparison of maximum power variations

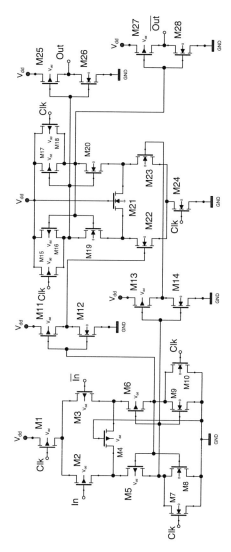

Figure B.4: SABL MS D-FF

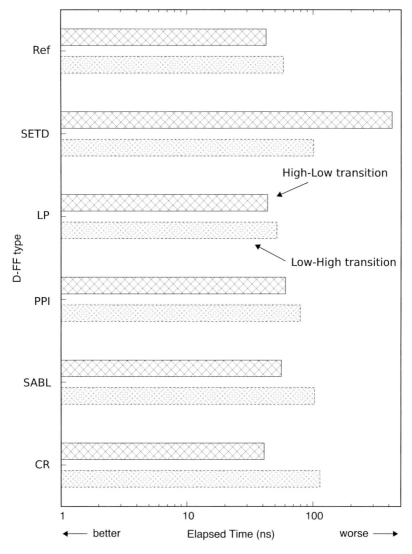

Figure B.5: Average consumed current $I_{V_{dd}}$

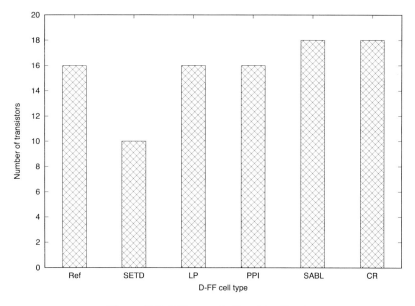

Figure B.6: Differences in Transistor Count

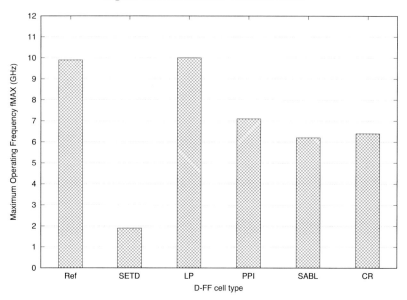

Figure B.7: Comparison Of Maximum Operating Frequencies

179

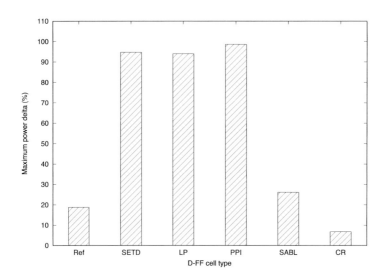

Figure B.8: Comparison of maximum power variations

Appendix C

Tristate Buffers - Additional Simulation Results

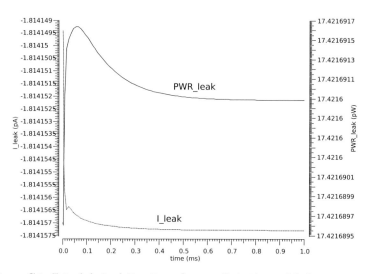

Figure C.1: Extended simulation time of power dissipation and leakage current in power-off mode with high Tox power gating transistors

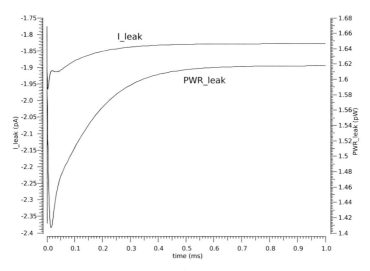

Figure C.2: Extended simulation time of power dissipation and leakage current in power-off mode of the modified low power tristate buffer

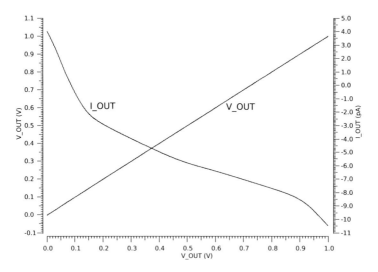

Figure C.3: Simulation result of modified low power tristate buffer without 'kink'

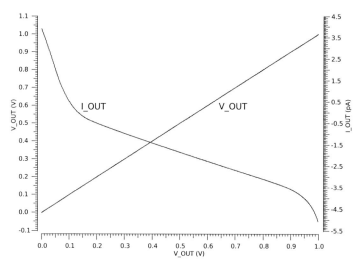

Figure C.4: Extended simulation time of power dissipation and leakage current in power-off mode of the modified low power tristate buffer

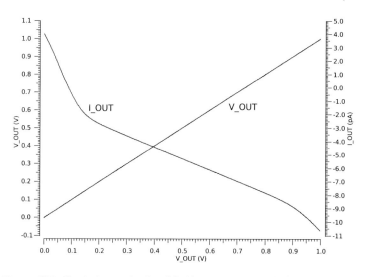

Figure C.5: Simulation result of modified low power tristate buffer without 'kink'

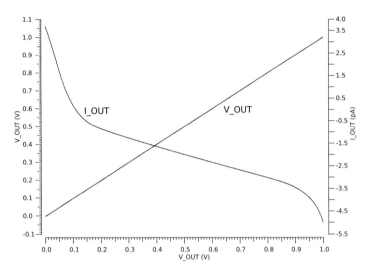

Figure C.6: Extended simulation time of power dissipation and leakage current in power-off mode of the modified low power tristate buffer

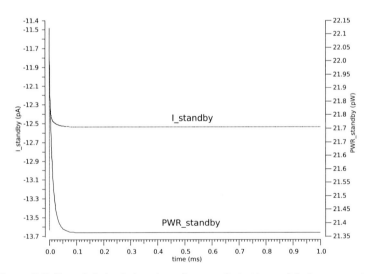

Figure C.7: Extended simulation time of power dissipation and leakage current in power-off mode of AT tristate buffer

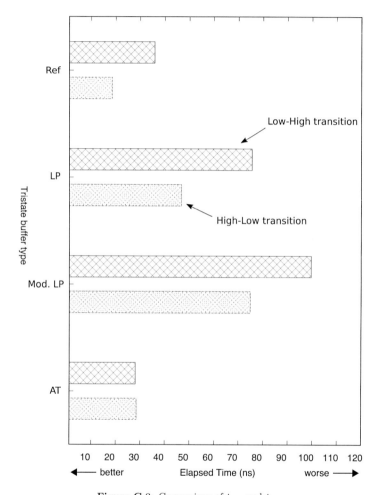

Figure C.8: Comparison of t_{HL} and t_{LH}

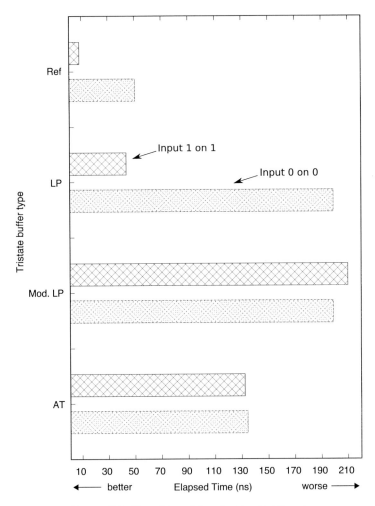

Figure C.9: Time delay while activating *highZ* mode

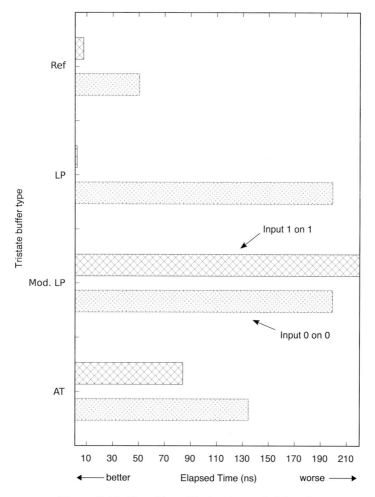

Figure C.10: Time delay while deactivating *highZ* mode

Appendix D

Partial SLICE Integration - Additional Figures and Simulation Results

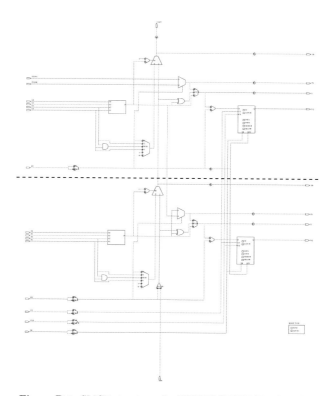

Figure D.1: SLICE structure of a *XILINX SPARTAN 3A* FPGA

189

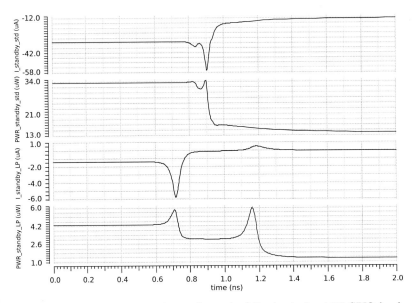

Figure D.2: Transient response in standby mode of the standard and LP GPIO (mod. LP tristate buffer)

List of Figures

195

Bibliography

[1] RABAEY, J.M.; CHANDRAKASAN, A.: *Digital integrated circuits - A design perspective* Prentice Hall 2004.

[2] ANDERSON, W.D.: *Designing with TTL integrated circuits* New York McGraw-Hill 1971.

[3] KUON, I; ROSE, J.: *Measuring the Gap Between FPGAs and ASICs* IEEE transactions on computer-aided design of integrated circuits and systems, Vol. 26, NO. 2, February 2007.

[4] XILINX, INC..: *XA Spartan-3A Automotive FPGA Family Data Sheet* Rev. 2.0, April 2011.

[5] ALTERA, CORP.: *Achieving Low Power in 65-nm Cyclone III FPGAs - White Paper* V 1.1, April 2007.

[6] INTEL, CORP.: *Intel Quartus Prime Design Software* Last accessed 2018-09-30.

[7] INTEL, CORP.: *Cyclone V Device Overview* CV-51001, 2018-05-07.

[8] ALTERA, CORP.: *Meeting the Low Power Imperative at 28 nm - White Paper* WP-01181-1-0, March 2012.

[9] INTEL, CORP.: *Achieving Lowest System Power with Low-Power 28-nm FPGAs - White Paper* WP-01158-2.1, September 2012.

[10] INTEL, CORP.: *Designing for Stratix 10 Devices with Power in Mind* AN-767, 2016-06-14.

[11] INTEL, CORP.: *Differences Among Intel SoC Device Families* UF-1005, 2018-08-22.

[12] MICROSEMI, CORP.: *Microsemi PolarFire FPGA Family - White Paper* Revision 1.0, October 2017.

[13] MICROSEMI, CORP.: *Product Overview PolarFire FPGA* Revision 1.0, October 2017.

[14] LATTICE, CORP.: *iCE40 UltraLiteTM Family Data Sheet* DS1051 Version 1.1, March 2015.

[15] LATTICE, CORP.: *iCE40 LP/HX Family Data Sheet* DS1040 Version 3.2, October 2015.

[16] LATTICE, CORP.: *iCE40LM Family Data Sheet* DS1045 Version 1.5, March 2015.

[17] LATTICE, CORP.: *iCE40 Ultra Wearable Development Platform User Guide* EB100 Version 1.0, July 2015.

[18] QUICKLOGIC, CORP.: *ArcticProTM Ultra-Low Power Embedded FPGA IP Product Brief*, 2017-04-04.

[19] XILINX, INC..: *BALANCING COST, POWER, AND PERFORMANCE FOR I/O CONNECTIVITY* (ver2.0) October 2011.

[20] XILINX, INC..: *Spartan-6 FPGAs: Performance, Power, and I/O Optimized for Cost-Sensitive Applications* WP396 v1.2, 2017-12-12.

[21] XILINX, CORP.: *Power Consumption at 45nm*, WP298 v2.0, 2016-08-08.

[22] XILINX, INC..: *Spartan-3A FPGA Family: Introduction and Ordering Information* DS529-1 v2.0, 2010-08-19.

[23] XILINX, INC..: *Total Power Advantage Using Spartan-7 FPGAs* WP488 v1.0, 2017-02-13.

[24] XILINX, INC..: *DS160 - Spartan-6 Family Overview* (ver2.0) October 2011.

[25] XILINX, CORP.: *Artix-7 FPGAs: Performance and Bandwidth in a Cost-Optimized Device*, WP423 v2.5.1, 2018-03-16.

[26] XILINX, CORP.: *UltraScale Architecture Low Power Technology Overview*, WP451 v1.1, 2015-10-15.

[27] XILINX, CORP.: *Lowering Power at 28 nm with Xilinx 7 Series Devices*, WP389 v1.3, 2015-01-05.

[28] XILINX, CORP.: *XILINX 7 SERIES FPGAS: BREAKTHROUGH POWER AND PER-FORMANCE, DRAMATICALLY REDUCED DEVELOPMENT TIME*, PN 2475-4, 2012.

[29] XILINX, CORP.: *7 Series FPGAs Data Sheet: Overview*, DS180 v2.6, 2018-02-27.

[30] QI, H.; AYORINDE, O.; CALHOUN, B.: *An ultra-low-power FPGA for IoT applications* IEEE SOI-3D-Subthreshold Microelectronics Technology Unified Conference (S3S), Burlingame, CA, USA, 16 - 19 October 2017.

[31] SUBBAREDDY, T.; REDDY, B.; UPADHYAY, H.: *Low power look-up table topologies for FPGAs* International Conference on Control, Instrumentation, Communication and Computational Technologies (ICCICCT), Kanyakumari, India, 10 - 11 July 2014.

[32] KUMAR, D.; KUMAR, P.; PATTANAIK, M.: *Performance Analysis of 90nm Look Up Table (LUT) for Low Power Application* 13th Euromicro Conference on Digital System Design: Architectures, Methods and Tools, Lille, France, 1 - 3 September 2010.

[33] SUZUKI, D.; HANYU, T.: *A low-power MTJ-based nonvolatile FPGA using self-terminated logic-in-memory structure* 26th International Conference on Field Programmable Logic and Applications (FPL), Lausanne, Switzerland, 29 August - 2 September 2016.

[34] WU, L.; ZHANG, G.; ZHAO, Y.: *A low power and radiation-tolerant FPGA implemented in FD SOI process* IEEE SOI-3D-Subthreshold Microelectronics Technology Unified Conference

(S3S), Monterey, CA, USA, 7 - 10 October 2013.

[35] QIAN, Z.; MARGALA, M.: *Low power RAM-based hierarchical CAM on FPGA* International Conference on ReConFigurable Computing and FPGAs (ReConFig14), Cancun, Mexiko, 8 - 10 December 2014.

[36] GROSSMANN, P.; LEESER, M.; ONABAJO, M.: *Minimum Energy Analysis and Experimental Verification of a Latch-Based Subthreshold FPGA* IEEE Transactions on Circuits and Systems II: Express Briefs, Volume: 59, Issue: 12, pp 942 - 946, December 2012.

[37] CHO, K.; LEE, S.; LEE, C.: *Low power multi-context look-up table (LUT) using spin-torque transfer magnetic RAM for non-volatile FPGA* International SoC Design Conference (ISOCC), Seoul, South Korea, 5 - 8 November 2017.

[38] YUAN, Z.; LIU, Y.; LI, JINYANG.: *CP-FPGA: Energy-Efficient Nonvolatile FPGA With Offline/Online Checkpointing Optimization* IEEE Transactions on Very Large Scale Integration (VLSI) Systems, Volume: 25, Issue: 7, July 2017.

[39] MUTUSAMY, S.; SUBRAMANIAM, V.: *Low-power dissipation using FPGA architecture* International Conference on Devices, Circuits and Systems (ICDCS), Coimbatore, India, 15 - 16 March 2012.

[40] CICCARELLI, L.; LODI, A.; CANEGALLO, R.: *Low Leakage Circuit Design for FPGAs* Proceedings of the IEEE 2004 Custom Integrated Circuits Conference, Orlando, FL, USA, 6 October 2004.

[41] PRIADARSHINI, A.; JAGADEESWARI, M.: *Low power reconfigurable FPGA based on SRAM* International Conference on Computer Communication and Informatics, Coimbatore, India, 4 - 6 January 2013.

[42] TUAN, T.; KAO, S.; RAHMAN, A.; DAS, S.; TRIMBERGER, S.: *A 90nm Low-Power FPGA for Battery-Powered Applications* FPGA' 06, Monterey, CA, USA, 22. - 25. June 2008.

[43] RAHMAN, A.; DAS, S.; TUAN, T.; TRIMBERGER, S.: *Determination of Power Gating Granularity for FPGA Fabric* IEEE 2006 Custom Intergrated Circuits Conference (CICC),

San Jose, CA, USA, 10. - 13. September 2006.

[44] LAMOUREUX, J.; LUK, W.: *An Overview of Low-Power Techniques for Field-Programmable Gate Arrays* NASA/ESA Conference in Adaptive Hardware and Systems, 22. - 24. February 2006.

[45] LI, F.; LIN, Y.; HE, L.: *Field Programmability of Supply Voltages for FPGA Power Reduction* IEEE Transactions on computer-aided design of integrated circuits and systems, VOL. 26, NO. 4, April 2007.

[46] ANDERSON, J.; NAJM, F.: *Active Leakage Power Optimization for FPGAs* IEEE Transactions on computer-aided design of integrated circuits and systems, VOL. 25, NO. 3, MARCH 2006.

[47] CHOW, C.; TSUI, L.; LEONG, P.; LUK, W.; WILTON, S.: *Dynamic Voltage Scaling for Commercial FPGAs* Proceedings. 2005 IEEE International Conference on Field-Programmable Technology, 11 - 14 December 2005.

[48] LODI, A.; CICCARELLI, L.; LOPARCO, D.; CANEGALLO, R.; GUERRIERI, R.: *Low Leakage Design of LUT-based FPGAs* Proceedings of ESSCIRC, Grenoble, France 2005.

[49] ANDERSON, J.; NAJM, F..: *A Novel Low-Power FPGA Routing Switch* Proceedings of the IEEE 2004 Custom Integrated Circuits Conference, Orlando, FL, USA, 22 November 2004.

[50] KUSSE, E.; RABAEY, J.: *Low-Energy Embedded FPGA Structures* ISLPED98, Monterey, CA, USA, April 1998.

[51] ISHIHARA, S.; HARIYAMA, M.; KAMEYAMA, M.: *A Low-Power FPGA Based on Autonomous Fine-Grain Power Gating* IEEE Transactions on Very Large Scale Integration (VLSI) Systems, 10 June 1998.

[52] GAYASEN, K.; VIJAYKRISHNAN, N.; KANDEMIR, M.; IRWIN, M.; TUAN, T.: *A Dual-VDD Low Power FPGA Architecture* International Conference on Field Programmable Logic and Applications FPL 2004, pp 145-157, 2004.

[53] HUSA, S.; MALLICK, M.; ANDERSON, J.: *Clock Gating Architectures for FPGA Power Reduction* International Conference on Field Programmable Logic and Applications, Prague, Czech Republic, 31 August - 02 September 2009.

[54] ZHOU, Y.; THEKKEL, S.; BHUNIA, S.: *Low power FPGA design using hybrid CMOS-NEMS approach* Proceedings of the 2007 international symposium on Low power electronics and design (ISLPED '07), Portland, OR, USA, 27 - 29 August 2007.

[55] CHEN, W.; LI, L.; LU, P.: *Design of FPGA's high-speed and low-power programmable interconnect* 13th IEEE International Conference on Solid-State and Integrated Circuit Technology (ICSICT), Hangzhou, China, 25-28 October 2016.

[56] KUMAR, H.; KARIYAPPA, B.: *Analysis of Low Power 7T SRAM Cell Employing Improved SVL (ISVL) Technique* International Conference on Electrical, Electronics, Communication, Computer and Optimization Techniques (ICEECCOT), Mysuru, India, 15-16 December 2017.

[57] REDDY, T.; MADAVI, B.: *Designing of Schmitt triggered-based architecture 8T SRAM of 256 bit cells under 45 nm technology for low power applications* International Conference on Intelligent Computing and Control (I2C2), Coimbatore, India, 23-24 June 2017.

[58] KIM, Y.; TONG, Q.; CHOI, K.: *Novel 8-T CNFET SRAM Cell Design for the Future Ultra-low Power Microelectronics* International SoC Design Conference (ISOCC), Jeju, South Korea, 3-6 November 2016.

[59] VO, H.: *A Double Regulated Footer And Header Voltage Technique For Ultra-Low Power IoT SRAM* IEEE 4th World Forum on Internet of Things (WF-IoT), Singapore, Singapore, 5-8 February 2018.

[60] BAGHEL, V.; AKASHE, S.: *Low power Memristor Based 7T SRAM Using MTCMOS Technique* Fifth International Conference on Advanced Computing & Communication Technologies, Haryana, India, 21-22 February 2015.

[61] SAYEED, A.; ALAM, N.; HASAN, M.: *Robust TFET SRAM cell for ultra-low power IoT application* International Conference on Electron Devices and Solid-State Circuits (EDSSC), Hsinchu, Taiwan, 18-20 October 2017.

[62] OHJA, S.; SINGH, O.; MISHRA, G.: *Analysis and design of single ended SRAM cell for low-power operation* 11th International Conference on Industrial and Information Systems (ICIIS), Roorkee, India, 3-4 December 2016.

[63] LIN, T.; ZHANG, W.; JHA, N.: *SRAM-Based NATURE: A Dynamically Reconfigurable FPGA Based on 10T Low-Power SRAMs* IEEE Transactions on Very Large Scale Integration (VLSI) Systems, Volume: 20 , Issue: 11 , November 2012.

[64] WANG, B.; ZHOU, J.; KIM, T.: *Ultra-low power 12T dual port SRAM for hardware accelerators* International SoC Design Conference (ISOCC), Jeju, South Korea, 3-6 November 2014.

[65] SELVAM, R.; SENTHILPARI, C.; LINI, L.: *Low power and low voltage SRAM design for LDPC codes hardware applications* IEEE International Conference on Semiconductor Electronics (ICSE2014), Kuala Lumpur, Malaysia, 27-29 August 2014.

[66] EVANGELENE, H.; SARMA, R.: *A novel low power hybrid flipflop using sleepy stack inverter pair* Science and Information Conference, London, UK, 27-29 August 2014.

[67] AGRAWAL, R.; PANDEY, J.; GUPTA, K.: *Implementation of PFSCL razor flipflop* International Conference on Computing Methodologies and Communication (ICCMC), Erode, India, 18-19 July 2017.

[68] JOSHI, P.; KHANDELWAL, S.; AKASHE, S.: *Implementation of Low Power Flip Flop Design in Nanometer Regime* Fifth International Conference on Advanced Computing & Communication Technologies, Haryana, India, 21-22 February 2015.

[69] NAKABAYASHI, T.; SASAKI, T.; OHNO, K.: *Low power semi-static TSPC D-FFs using split-output latch* International SoC Design Conference, Jeju, South Korea, 17-18 November 2011.

[70] DEMASSA, T.; CICCONE, Z.: *Digital Integrated Circuits* John Wiley & Sons, Inc., 1996.

[71] SKAHILL, K.: *VHDL for Programmable Logic* ADDISON-WESLEY Publishing Company, 1996.

[72] BROWN, S.; FRANCIS, R.; ROSE, J.; VRANESIC, Z.: *Field-Programmable Gate Array* Kluwer Academic Publishers, 1993.

[73] BAKOGLU, H.: *Circuits, Interconnections and Packaging for VLSI* ADDISON-WESLEY Publishing Company, 1990.

[74] AUER, A.; RUDOLF, D.: *FPGA Feldprogrammierbare Gate Arrays* Hüthig Buch Verlag, Heidelberg, Germany, 1995.

[75] JAEGER, R.; BLALOCK, T.: *Microelectronic Circuit Design* The McGraw-Hill Companies, 1997.

[76] EISENBARTH, T.: *Cryptography and cryptanalysis for embedded systems*, PhD thesis, Europäischer Universitätsverlag, Bochum, 2010.

[77] KOCHER, P.; JAFFE, J.; JUN, B.: *Differential Power Analysis* CRYPTO '99: Proceedings of the 19th Annual International Cryptography Conference on Advances in Cryptology, pp. 388–397, London, UK, Springer-Verlag, 1999.

[78] STANDAERT, F.-X.; ARCHAMBEAU, C.: *Using Subspace-Based Template Attacks to Compare and Combine Power and Electromagnetic Information Leakages* CHES 2008: Cryptographic Hardware and Embedded Systems, pp. 411–425, Springer, 2008.

[79] CORON, J.: *Resistance against Differential Power Analysis for Elliptic Curve Cryptosystems* Cryptographic Hardware and Embedded Systems - CHES 1999, pp. 292–302, 1999.

[80] BRIER, E.; CLAVIER, C.; OLIVIER, F.: *Correlation Power Analysis with a Leakage Model* Cryptographic Hardware and Embedded Systems - CHES 2004, pp. 16–29, 2004.

[81] PROUFF, E.; RIVAIN, M.; BEVAN, R.: *Statistical Analysis of Second Order Differential Power Analysis* IEEE Transactions on Computers, pp. 799–811, 2009.

[82] CHEN, Z.; CHEN, Z.: *Dual-Rail Random Switching Logic: A Countermeasure to Reduce Side Channel Leakage* Cryptographic Hardware and Embedded Systems - CHES 2006, pp.

242–254, 2006.

[83] POPP, T.; MANGARD, S.: *Masked Dual-Rail Pre-charge Logic: DPA-Resistance Without Routing Constraints* Cryptographic Hardware and Embedded Systems CHES 2005, pp. 172–186, 2005.

[84] TIRI, K.; VERBAUWHEDE, I.: *A logic level design methodology for a secure DPA resistant ASIC or FPGA implementation* Proceedings Design, Automation and Test in Europe Conference and Exhibition, Paris, France, 16. - 20. February 2004.

[85] PIGUET, C.: *Low-power processors and systems on chips* CRC Press, 2005.

[86] PIGUET, C.: *Logic Design for Low-voltage/Low-power CMOS Circuits* CRC Press, 2005.

[87] MALONE, M.S.: *The Microprocessor: A Biography* Springer Verlag 1995.

[88] ACHLEITNER, S.: *Differential Power-Analysis Attacks: A Practical Example for Hardware Countermeasures Protecting Cryptographic Circuits* VDM Verlag, 2008.

[89] FÜRST, S.: *Challenges in the Design of Automotive Software* European Design and Automation Association (2010), pp. 256–258.
$http : //dl.acm.org/citation.cfm?id = 1870926.1870987$

[90] HUEBNER, M.; SCHUCK, C.; KIIHNLE, M.; BECKER, J.: *New 2-Dimensional Partial Dynamic Reconfiguration Techniques for Real-time Adaptive Microelectronic Circuits* Proceedings of the 2006 Emerging VLSI Technologies and Architectures (ISVLSI?06), 2006.

[91] FEI, L.; CHEN, D.; HE, L.; CONG, J.: *Architecture Evaluation for Power-Efficient FPGAs* FPGA'03, Monterey, California, USA, February 2006.

[92] FARINELLI, A.; ROGERS, A.; PETCU, A.; JENNINGS, N.: *Decentralised Coordination of Low-power Embedded Devices Using the Max-sum Algorithm* Proceedings of the 7th International Joint Conference on Autonomous Agents and Multiagent Systems - Volume 2, pp. 639–646, 12. - 16. May 2008.

[93] KULKARNI, S.; SRIVASTAVA, A.; SYLVESTER, D.: *A New Algorithm for Improved VDD Assignment in Low Power Dual VDD Systems* Proceedings of the 2004 International Symposium on Low Power Electronics and Design, pp. 200–205, 9. - 11. August 2008.

[94] SINGH, J.; MOHANTY, S.; PRADHAN, D.: *Robust SRAM Designs and Analysis* Springer Verlag 2014.

[95] GREMZOW, C.: *Entwurf digitaler Systeme in VHDL* Chair of Automation and Computer Science, University of Wuppertal 2016.

[96] SZE, S.; LEE, M.: *Semiconductor Devices - Physics and Technology 3rd Edition* John Wiley & Sons Inc 2013.

[97] PAL, A.: *Low-Power VLSI Circuits and Systems* Springer India 2015.

[98] ITOH, K.: *VLSI Memory Chip Design* Springer India 2015.

[99] TIETZE, U.; SCHENK, C.: *Halbleiter-Schaltungstechnik* Heidelberg Springer Vieweg 201g.

[100] SARKAR, A.; DE, S.; CHANDA, M.; SARKAR, C.: *LOW POWER VLSI DESIGN* Springer 2001.

[101] CARLSON, I.; ANDERSON, S.: *A high density, low leakage, 5T SRAM for embedded caches* Solid-State Circuits Conference, 2004. ESSCIRC 2004, pp. 215–218, September 2004.

[102] HUEBNER, M.; PAULSSON, K.; BECKER, J.: *Parallel and Flexible Multiprocessor System-On-Chip for Adaptive Automotive Applications based on Xilinx MicroBlaze Soft-Cores* Proceedings of the 19th IEEE International Parallel and Distributed Processing Symposium (IPDPS?05), 2005.

[103] KHURRAM, M.; KUMAR, H.; CHANDAK, A. SARWADE, V.: *Enhancing connected car adoption: Security and over the air update framework* 2016 IEEE 3rd World Forum on Internet of Things (WF-IoT), pp. 194–198, 12. - 14. December 2016.

[104] INGERMAN, P. Z.: *The Dangers of Distributed Intelligence [Perspective]* IEEE Technology and Society Magazine, Volume: 36, Issue: 2, June 2017.

[105] XILINX, INC.: *XA Spartan-3A Automotive FPGA Family Data Sheet*, Xilinx Product Specification, Rev. 2.0, April 2011.

[106] ANDERSON, J.; NAJM, F.: *Active Leakage Power Optimization for FPGAs* IEEE Transactions on computer-aided design of integrated circuits and systems, Vol. 25, NO.3, March 2006.

[107] INFINEON, AG.: *Halbleiter*, Publicis Corporate Publishing 2004.

[108] ROY, K.: *Low power CMOS VLSI circuit design* John Wiley & Sons Inc 2000.

[109] HAUSNER, J.: *Integrated Digital Circuits* Chair of Integrated Systems, Universtiy of Bochum 2011.

[110] NODA, K.; TAKEDA, K.; MATSUI, K.; NAKAMURA, N.: *A loadless CMOS four-transistor SRAM cell in a 0.18- μm logic technology* IEEE Transactions on Electron Devices, December 2001.

[111] ALY, R. E.; FAISAL, M.; BAYOUMI, M.: *Novel 7T SRAM cell for low power cache design* Proceedings 2005 IEEE International SOC Conference, September 2005.

[112] MAXFIELD, C.: *The Design Warrior's Guide to FPGAs: Devices, Tools and Flows* Newnes 2004.

[113] SHARMA, M.; NOOR, A.; TIWARI, S.; SINGH, K.: *An Area and Power Efficient Design of Single Edge Triggered D-FlipFlop* 2009 International Conference on Advances in Recent Technologies in Communication and Computing, pp. 478–481, October 2009.

[114] KO, U.; BALSARA, P.: *High-performance energy-efficient D-flip-flop circuits* IEEE Transactions on Very Large Scale Integration (VLSI) Systems, February 2000.

[115] SCHNEIER, B.: *Applied Cryptography: Protocols, Algorithms and Source Code in C 20th Anniversary Edition* John Wiley & Sons Inc 2015.

[116] BELLAOUAR, A.; ELMASRY, M.: *Low-Power Digital VLSI Design Circuits and Systems* Kluwer Academic Publishers 1995.

[117] PAVLOV, A.: *Design and Test of Embedded SRAMs* PhD thesis, University of Waterloo, Ontario, May 2005.

[118] MANGARD, S.; OSWALD, E.; POPP, T.: *Revealing the Secrets of Smart Cards* Springer Science+Business Media, 2007.

[119] PAVLOV, A.: *Side-Channel Analysis Aspects of Lightweight Block Ciphers* Diploma thesis, Ruhr-University of Bochum, Bochum, 2009.

[120] UYEMURA, J.: *CMOS Logic Circuit Design* Kluwer Academic Publishers 1999.

OWN PUBLICATIONS

[121] NIEWIADOMSKI, K.; GREMZOW, C.; TUTSCH, D.: *4T Loadless SRAMs for Low Power FPGA LUT Optimization* Proceedings of the 9th International Conference on Adaptive and Self-Adaptive Systems and Applications (ADAPTIVE 2017), pp. 1–7, February 2017.

[122] NIEWIADOMSKI, K.; TUTSCH, D.: *Low Power Charge Recycling D-FF* The Tenth International Conference on Advances in Circuits, Electronics and Micro-electronics (CENICS 2017), pp. 1–6, September 2017.

[123] NIEWIADOMSKI, K.; TUTSCH, D.: *Low Power Tristate Buffer for Mobile Applications* Proceedings of the 10th International Conference on Adaptive and Self-Adaptive Systems and Applications (ADAPTIVE 2018), pp. 1–6, February 2018.

[124] NIEWIADOMSKI, K.; TUTSCH, D.: *Enhanced 4T Loadless SRAM Comparison With Selected Volatile Memory Cells* International Journal on Advances in Systems and Measurements, IARIA, Vol. 10, No. 3&4, pp. 139–149, December 2017.

[125] Niewiadomski, K.; Tutsch, D.: *Low Power Optimized and DPA Resistant D-FF for Versatile Mobile Applications* International Journal on Advances in Systems and Measurements, IARIA, Vol. 11, No. 1&2, pp. 100–110, June 2018.

[126] Niewiadomski, K.; Tutsch, D.: *Newly Developed Low Power Tristate Buffers for Low-power and High Performance Applications with Limited Energy Resources* International Journal on Advances in Systems and Measurements, IARIA, accepted for publication.